SpringerBriefs in Applied Sciences and Technology

SpringerBriefs present concise summaries of cutting-edge research and practical applications across a wide spectrum of fields. Featuring compact volumes of 50 to 125 pages, the series covers a range of content from professional to academic.

Typical publications can be:

- A timely report of state-of-the art methods
- An introduction to or a manual for the application of mathematical or computer techniques
- A bridge between new research results, as published in journal articles
- A snapshot of a hot or emerging topic
- An in-depth case study
- A presentation of core concepts that students must understand in order to make independent contributions

SpringerBriefs are characterized by fast, global electronic dissemination, standard publishing contracts, standardized manuscript preparation and formatting guidelines, and expedited production schedules.

On the one hand, **SpringerBriefs in Applied Sciences and Technology** are devoted to the publication of fundamentals and applications within the different classical engineering disciplines as well as in interdisciplinary fields that recently emerged between these areas. On the other hand, as the boundary separating fundamental research and applied technology is more and more dissolving, this series is particularly open to trans-disciplinary topics between fundamental science and engineering.

Indexed by EI-Compendex, SCOPUS and Springerlink.

More information about this series at http://www.springer.com/series/8884

Mohd Haziman Bin Wan Ibrahim ·
Shahiron Shahidan · Hassan Amer Algaifi ·
Ahmad Farhan Bin Hamzah ·
Ramadhansyah Putra Jaya

Properties of Self-Compacting Concrete with Coal Bottom Ash Under Aggressive Environments

Springer

Mohd Haziman Bin Wan Ibrahim
Faculty of Civil Engineering and Built
Environment
Jamilus Research Centre, Universiti Tun
Hussein Onn Malaysia, Batu Pahat Johor,
Malaysia

Shahiron Shahidan
Faculty of Civil Engineering and Built
Environment
Jamilus Research Centre, Universiti Tun
Hussein Onn Malaysia, Batu Pahat Johor,
Malaysia

Hassan Amer Algaifi
Faculty of Civil Engineering and Built
Environment
Jamilus Research Centre, Universiti Tun
Hussein Onn Malaysia, Batu Pahat Johor,
Malaysia

Ahmad Farhan Bin Hamzah
National Hydraulic Research Institute
of Malaysia (NAHRIM)
Selangor, Malaysia

Ramadhansyah Putra Jaya
Civil Engineering Department
College of Engineering
Universiti Malaysia Pahang
Kuantan, Pahang, Malaysia

ISSN 2191-530X   ISSN 2191-5318   (electronic)
SpringerBriefs in Applied Sciences and Technology
ISBN 978-981-16-2394-3   ISBN 978-981-16-2395-0   (eBook)
https://doi.org/10.1007/978-981-16-2395-0

This Springer imprint is published by the registered company Springer Nature Singapore Pte Ltd.
The registered company address is: 152 Beach Road, #21-01/04 Gateway East, Singapore 189721,
Singapore

Introduction

Concrete is one of the commonly used construction materials, known for its high compression strength. Although considered strong, concrete is a brittle material, weak in tension, promoting the development and prorogation of microcracks. Concrete also contains numerous pores and undesirable chemical ions, such as calcium hydroxide, specifically at the interfacial transition zone between aggregate and cement paste, which is regarded as a weakness in concrete. These conditions permit the entry of harmful ions, such as sulfate, chloride, and seawater, that penetrate the concrete cover and thereby accelerate the deterioration of the concrete structure. Therefore, concrete life span has emerged as hotspot research in the construction industry, specifically centered around concrete structures are exposed to aggressive environments.

This book intends to provide insight into the influence of coal bottom ash on the performance of self-compacting concrete exposed to aggressive conditions. The target audiences of the book are academicians, researchers, and students who seek in-depth information on the performance and properties of self-compacting concrete with bottom ash exposed to tap water, sulfate, chloride, and seawater. The book features information on vital topics essential to ensure the sustainability of the concrete structure. The entire book is divided into four chapters, with each chapter presented with tables and graphics to encourage better understanding and visualization of this topic. A list of references is also put together for further information.

Chapter 1 presents the background information on concrete deterioration. In particular, the concept of salt crystallization in the concrete matrix is briefly highlighted. Then, the influence of several factors affecting salt crystallization, such as type of cement, temperature, curing, water-cement ratio, and the presence of sulfate and chloride, are also presented. The aggressive exposure conditions, such as wetting and drying, freeze and thaw, are also discussed in detail. These conditions favour the movement of the harmful ions into concrete and accelerate the degradation in the concrete.

Chapter 2 presents a brief background on the literature on using coal bottom ash to improve concrete properties. The effects of adding local coal bottom ash as a fine aggregate replacement on the fresh and hardened properties of self-compacting concrete exposed to tap water are presented in detail. For fresh properties, flowability, passing ability, and segregation resistance test are taken into account. For the hardened properties, unit weight, compressive strength, tensile strength, flexural strength, permeable pore space, and water absorption at different interval times are evaluated. The microstructural analyses of self-compacting concrete incorporating coal bottom ash (SCC-BA) performed by Scanning Electron Microscopy (SEM) and X-ray diffraction (XRD) are also presented.

Chapter 3 elucidates the impact and extent to which coal bottom ash could affect the performance of self-compacting concrete exposed to sulfate and chloride attack. It started with an explanation of basic understanding of the mechanism of both sulfate and chloride attack on concrete. Then, the evolution of compressive strength of SCC-BA exposed to both sulfate and chloride attack age (7, 28, 60, 90, and 180 days) is evaluated and discussed. The inferences from the rapid chloride-ion permeability test, chloride penetration by rapid migration test, and carbonation depth test are also highlighted and discussed. The examination of microstructure of SCC-BA exposed to both sulfate and chloride attack using SEM and XRD is presented.

Chapter 4 provides detailed information on the performance of SCC-BA exposed to seawater at different amounts of coal bottom ash. In particular, it highlights the effect of CBA on the compressive strength including the value of strength reduction and weight loss due to seawater. After that, the durability aspect of SCC-BA exposed to seawater is also presented and discussed including chloride-ion permeability test, chloride penetration by rapid migration test, and carbonation depth test for different ageing periods (7, 28, 60, 90, and 180 days). SEM and XRD are also presented to evaluate the SCC-BA microstructure.

The authors are grateful to all the reviewers for their time and constructive comments. With best wishes.

Contents

Chapter 1
Introduction

Abstract Concrete durability remains a concern in the construction industry due to salt attacks that accelerate concrete deterioration. The salt attack has a significant impact on concrete durability and thus reduces the concrete lifespan. Detailed background information on concrete deterioration is presented in this chapter. This chapter also covers the concepts and mechanisms of salt attack in concrete. The influential factors affecting salt attack such as cement type, water binder ratio, and curing condition are discussed in detail. Moreover, aggressive exposure conditions such as wetting and drying and freeze and thawing are also briefed in this chapter.

1.1 Historical Background of Concrete Deterioration

Concrete is the most common material used in construction for various purposes such as dams, highways, buildings, foundations, etc. Besides, concrete is fundamentally brittle and thus is susceptible to form cracks, which then affects the durability of concrete structures. In other words, there is a strong tendency for concrete to form cracks. This is unsurprising, as the concrete matrix has been regarded for its brittleness and heterogeneity of the material. These cracks can be created even without any external load wherein they originate from the early age thermal gradient, bleeding, shrinkage, and expensive reaction (Algaifi et al. 2020). In addition, both microcracks and porosity could generate interconnected flow paths inside the concrete matrix in the presence of an external load. This fact could facilitate the ingress of aggressive material and deteriorate the concrete structure (Algaifi et al. 2018). The most serious concern is the premature deterioration of concrete exposed to aggressive exposure environments such as salt attacks, classified into salt crystallization, seawater attack, sulphate attack, and chloride attack. This is because the government has allocated a considerable budget to repair the existing deteriorated concrete structures due to the corrosion of the reinforcement.

From another point of view, concrete deterioration is also significantly affected by curing conditions. For example, deeper penetration of harmful ions in the concrete matrix could be accelerated when the concrete is subjected to cyclic wetting and

© The Author(s), under exclusive license to Springer Nature Singapore Pte Ltd. 2021 1
M. H. Bin Wan Ibrahim et al., *Properties of Self-Compacting Concrete
with Coal Bottom Ash Under Aggressive Environments*, SpringerBriefs in Applied
Sciences and Technology, https://doi.org/10.1007/978-981-16-2395-0_1

drying conditions. Furthermore, as well known, the penetration rate of concrete exposed to a wetting–drying environment is time-dependent. By increasing the period of wetting–drying, the rate of ion penetration also increases.

Consequently, both the codes and specifications of concrete construction involving concrete exposed to harmful environments (sulphate, chloride, sea water, and so on) focussed on concrete mixture quality such as minimum concrete cover thickness, cement content, and water to cement ratio (w/c). For instance, a minimum thickness cover of 50 mm, a maximum w/c ratio of 0.45, and a minimum cement content of 400 kg/m^3 were recommended by American Concrete Institute (ACI) Committee 357 on the Design and Construction of Fixed Concrete Offshore Structures for marine structures (ACI Committee 357 1994). In the same context, both the European Codes (ENV-206 1992) and the British Standards (BS 8110 1985) have similar maritime structure requirements.

As expected, the improvement of the fresh and hardened properties of self-compacting concrete (SCC) was higher than that of conventional concrete. In other words, the development of self-compacting concrete was to address the need for improved workability and durability. On the other hand, coal bottom ash (CBA or BA), which comprises particles similar to natural size, is also recognized as a partial replacement of fine aggregates in the concrete matrix. In this study, the incorporation of CBA in SCC was investigated and evaluated. In particular, the performance of SCC-BA exposed to tap water and aggressive conditions is evaluated and discussed further in the following chapters, while the mechanisms and factors affecting the concrete deterioration, such as a salt attack, are highlighted in the following sections.

1.2   Salt Attack

The salt attack is considered to be a severe mechanism that would negatively deteriorate the concrete structure. However, by producing more durable concrete, it could be considered a controlled phenomenon. More prolonged exposure to salt attacks results in the crystallization of salt inside concrete pores. This result causes the disintegration of structure and concrete matrix. The salt attack's basic mechanism is the penetration of the salt ions into the concrete matrix in the presence of water. Salt residues are deposited inside the concrete pores due to the water evaporation through the dryer area. The wetting–drying process causes a slow rise in salt in the capillary and concrete pores. As a result, the salt remains more concentrated, and crystals are formed at this time. During the growth of crystals, the presence of microparticles in mortars increases, and internal stresses and forces are exerted. As such, the internal force generated may cause damage.

1.2.1 Influential Factors Affecting Salt Attack

As discussed earlier, the salt attack is a complex problem that deteriorates the strength of concrete. It depends on several factors related to the exposure conditions and materials presented in the next section. From the viewpoint of exposure condition, the type of ion, ion concentration, and temperature play a major role in accelerating the salt attack mechanism. In terms of the material properties, curing condition, degree of hydration, water to cement ratio, type of cement, and mineral admixture are also important parameters that either increase or decrease the rate of concrete deterioration at the time of the salt attack.

1.2.1.1 Composition of Cement

The resistance of salt attack is either increased or decreased depending on the chemical composition of cement. For example, the sulphate attack depends on the amount of tricalcium aluminate (C_3A) and calcium hydroxide ($Ca(OH)_2$) that are present in the hydrated concrete. A cement would be a good sulphate resistant when the amount of C_3A is low. Therefore, the sulphate-resisting Portland cement would act as a good material to resist salt attack as it has low C_3A.

On the other hand, the amount of tricalcium silicate (C_3S) in modern cement is high, resulting in a high calcium hydroxide concentration. This fact promotes the susceptibility of a sulphate attack. To overcome this problem, various mineral mixtures are blended with cement, such as silica fume (SF), ground granulated blast furnace slag (GGBFS), and fly ash (FA). On using mineral admixture, calcium hydroxide is consumed due to the pozzolanic reaction and improves both the strength and durability of concrete. This is because calcium hydroxide in concrete exposed to aggressive chemical ions is considered a weakness in terms of durability aspect. Given the above scenario, the cement composition is an important parameter that would impact the resistance of the salt attack.

1.2.1.2 Curing Condition

Curing condition is another parameter that affects the compressive strength of the concrete and the performance and lifespan of the concrete structure. At an early stage, tap water curing is necessary to prevent the evaporation of the water required for hydration. Tap water curing is required to achieve optimum performance by completing the hydration process. In contrast, aggressive curing conditions such as sulphate, chloride, and seawater negatively impact concrete properties, specifically at a later age. The repeated wetting and drying cycles also have a negative impact on concrete durability, which is discussed in detail in the next chapters. It is inferred that the concrete mixture must be properly designed to withstand any aggressive exposure.

1.2.1.3   Water Binder Ratio

Water cement ratio is the main factor in the development of concrete strength. By decreasing the water–cement ratio, the compressive strength of concrete increased due to improved concrete microstructure. This is because the water–cement ratio decrease would minimize the pore space inside the concrete matrix and thus make the concrete denser.

1.2.1.4   Temperature of Exposure

A high-temperature environment has a negative impact on concrete behaviour. As temperatures increase, water and moisture begin to evaporate, and as a result, the integrity of the concrete matrix is weakened. Due to the high temperature, the gel water of hydrated cement particles would evaporate due to gel dehydration, causing damage to the concrete matrix. Moreover, cracks are initiated and gradually generated, leading to concrete spalling.

1.2.1.5   Presence of Sulphate

In the possibility of salt attack, the rate of concrete deterioration is also dependent on the type of sulphate and its concentration. Al-Amoudi (2002) studied the performance of concrete samples subjected to magnesium, sodium, and mixed sulphate solutions. The comparison of strength reduction was observed in these environments for 100 days of exposure. After that, the strength reduction significantly increased for concrete specimens exposed to magnesium sulphate compared to other solutions. Further expansion was observed in all mortar specimens subjected to the sodium sulphate environment. However, the blended cement showed superior resistance to sodium sulphate solution. The concrete deterioration due to sulphate attack also depends on sulphate concentration. By increasing the sulphate concentration, the rate of concrete deterioration also increased.

1.2.1.6   Presence of Chloride

Both chloride and sulphate ions have a negative effect on the concrete lifespan. In the case of chloride ions, cement hydrates react with chloride ions to form the Freidel salt, which has no harmful effect on concrete. However, the protective layer of steel reinforcement would be destroyed if the chloride contents were higher than the threshold value. As such, the steel reinforcement gets corroded in the presence of chloride, moisture, and oxygen. Alternately, the concrete deterioration increased in the presence of both chloride and sulphate solution. (Al-Amoudi 2002) investigated the performance of the blended cement in a sulphate attack. Solutions (2.1% SO_4+ 15.7% Cl and 2.1% SO_4 solution) were studied for 365 days. The extension of specimens and reduction of strength were considered to represent and assess deterioration. Based on the results, the pure sulphate solution was more aggressive than the exposure of sulphate–chloride solution.

1.3   Aggressive Exposure Condition

Premature concrete deterioration is now a serious problem for concrete structures exposed to harsh environments. This is because concrete is regarded as a porous material with a complex pore structure system, i.e. microstructure. Being the critical focus of the research community, the microstructure could either positively or negatively affect the performance of concrete exposed to harmful environments in both the short and long terms. In particular, the important aspect of the microstructure is that it is composed of a complex pore system, such as pores and interconnectivity pores and invertible microcracks that facilitate the ingress of aggressive chemical ions into the concrete. In addition, concrete is a non-homogenous material composed of three phases, namely aggregate, cement paste, and the interfacial transition zone (ITZ), which is also regarded as a weak layer in the cement-based matrix. The ITZ contains high porosity, calcium hydroxide crystals, and microcracks. Both interconnectivity pore systems and invertible microcracks provide an easy path for aggressive material to attack the concrete and affect the lifespan of the concrete. For example, alkali–silica reactions, chloride attack, sulphate attack, and carbonation in concrete can influence the performance of concrete. Therefore, these issues concerning durability, aggregate mix proportions, and material selection should be well addressed.

1.3.1   Wetting and Drying

The process of wetting and drying cycles is supposed to be a harsh condition as it facilitates the movement of moisture involving harmful ions into the concrete pores. As such, the wetting and drying cycles would result in problems in the concrete and accelerate its degradation. This is- because destructive elements, such as chlorides, alkalis, and sulphates, could easily be moved and accumulated inside a concrete matrix exposed to wetting and drying cyclic. For example, chloride penetration in concrete pores is achieved through sequential steps, such as absorption, diffusion, and dispersion. The mechanism of chloride entry into the structures subjected to wet-drying cycles depends mainly on absorption and diffusion. Similarly, it is interesting to note that the corrosion of the steel reinforcement due to the chloride attack shortens the concrete lifespan.

An experiment was conducted by Ye et al. (2016) to investigate the chloride and carbonation penetration process in concrete. The concrete specimens were exposed to carbonation and cyclic drying–wetting. Depending on the concrete properties, various deteriorating interactive mechanisms have been established to describe the chloride penetration. In particular, it has been inferred that concrete is becoming more vulnerable to chloride attack on being incorporated with supplementary cementitious materials undergoing a combined deterioration of carbonation and cyclic drying–wetting process.

1.3.2 Freeze and Thaw

Salt scaling or internal frost damage may cause freezing and thaw damage. The provision of an adequate air void system would enhance the resistance to internal frost damage. It includes proper air void spacing, size, and total volume of air void. Resistance could also be improved by using supplementary cementitious materials. Supplementary cementitious materials aimed to minimize the water to cement ratio and subsequently reduce permeability and larger pores. An increase could also improve the resistance to internal frost damage in concrete strength.

On the other hand, the provision of adequate trained air void systems could prevent salt scaling. Resistance enhancement of the freeze–thaw can be achieved by adequate entrain air and low water–cement ratios of the material. The use of slag and fly ash may increase the resistance to salt scaling. Furthermore, self-compacting concrete demonstrated its ability to withstand freeze–thaw compared to conventional concrete. The reduction of water-powder and water–cement ratios in SCC mixtures has enhanced internal resistance to frost. However, some admixture may have an impact on the content of air and air void system characteristics. Khayat and Assaad (2002) demonstrated that the characteristics of air voids present in SCC were similar to conventional concrete. Also, by increasing the content of cementitious materials content, the stability of the air void increased.

1.4 Conclusion

It can be concluded that concrete durability has emerged to be an attractive research area in the concrete community, specifically focussing on the exposure of concrete to chemical attacks. A harsh conditions such as chloride and sulphate ions significantly affect the durability of concrete. Additionally, wetting and drying cycles as curing environments also have a negative impact on the concrete lifespan. The self-compacting concrete exhibited a positive result for minimizing or eliminating the deteriorating phenomena of concrete compared to that of the control mixture. Besides, there would be an interest in enhancing concrete durability by combining CBA and SCC benefits. Further, in the following chapters, self-SCC-BA subject to harsh environments and tap water is discussed in detail.

References

Al-Amoudi, O. 2002. Durability of plain and blended cements in marine environments. *Advances in Cement Research* 14: 89–100.

Algaifi, H. A., S. A. Bakar, A. R. M. Sam, A. R. Z. Abidin, S. Shahir, and W. A. H. Al-Towayti. 2018. Numerical modeling for crack self-healing concrete by microbial calcium carbonate. *Construction and Building Materials* 189: 816–824.

Algaifi, H. A., S. A. Bakar, A. R. M. Sam, M. Ismail, A. R. Z. Abidin, S. Shahir, and W. A. H. Altowayti. 2020. Insight into the role of microbial calcium carbonate and the factors involved in self-healing concrete. *Construction and Building Materials* 254: 119258.

Khayat, K. H., and J. Assaad. 2002. Air-void stability in self-consolidating concrete. *ACI Materials Journal* 99: 408–416.

Ye, H., X. Jin, C. Fu, N. Jin, Y. Xu, and T. Huang. 2016. Chloride penetration in concrete exposed to cyclic drying-wetting and carbonation. *Construction and Building Materials* 112: 457–463.

Chapter 2
CBA Self-compacting Concrete Exposed to Water Curing

Abstract The application of self-compacting concrete is recognized as a sustainable strategy for achieving high workability and improved durability. However, aggressive chemical ions would have a negative effect on the performance of self-compacting concrete. In harsh environments, the combination of self-compacting concrete and coal bottom ash could offer potential advantages of durability for concrete structures. In the present chapter, a detailed background of self-compacting concrete exposed to tap water is provided. An in-depth insight into the impact and extent to which the coal bottom ash could enhance both the fresh and hardened properties of SCC-BA exposed to normal water curing.

2.1 Historical Background of CBA Concrete

The usage of coal bottom ash in concrete has become an interesting research topic in the construction industry. This is because of the similarities in the size of the CBA particles and the nature of river sand. Moreover, CBA particles usually range from fine sand to fine gravel. Previously, researchers investigated CBA use in concrete as a partial replacement of cement or as fine aggregate. Table 2.1 presents the properties of concrete incorporation coal bottom ash as sand replacement.

It can be inferred that many researchers investigated and evaluated the properties of conventional concrete containing coal bottom ash and the effects on subjecting it to water curing. Besides, the improvement in the conventional concrete properties varied between each other based on the replacement percentage and properties of coal bottom ash. The influence of local coal bottom ash on the properties of self-compacting concrete exposed to tap water is evaluated and discussed in detail in the next section.

© The Author(s), under exclusive license to Springer Nature Singapore Pte Ltd. 2021
M. H. Bin Wan Ibrahim et al., *Properties of Self-Compacting Concrete with Coal Bottom Ash Under Aggressive Environments*, SpringerBriefs in Applied Sciences and Technology, https://doi.org/10.1007/978-981-16-2395-0_2

Table 2.1 Properties of CBA self-compacting concrete

Properties	Authors	Findings	Remarks
Fresh properties/workability of concrete	Singh and Siddique (2013) (Zainal Abidin et al. 2015)	The fine and irregular-shaped, rough-textured, and porous particles of coal bottom ash increase, thereby increasing the interparticle friction, hindering fresh concrete's flow characteristics	This is positive for self-compacting concrete as the flow characteristics need to be satisfied to comply with the EFNARC standards as self-compacting concrete
	Abubakar and Baharudin (2012)	The fresh concrete properties gradually reduced based on the amount of fine coal bottom ash	The incorporated coal bottom ash in concrete could be included with superplasticizer and viscosity modifying admixtures
Unit weight	Ibrahim et al. (2015)	The density of concrete specimens decreased with an increase of coal bottom ash level	The unit weight of coal bottom ash is comparatively low compared to sand and could be potentially utilized for lightweight concrete production
Compressive strength	Topçu et al. (2014)	Compressive strength of bottom ash concrete mixtures developed in a comparable state or higher to that of control concrete mixture	An improvement of compressive strength of concrete incorporating coal bottom ash was observed. The addition of chemical admixture helps to minimize water demand to achieve higher strength
Tensile strength	Aggarwal and Siddique (2014)	In the case of coal bottom ash as sand replacement in concrete, tensile strength decreases as it increases the coal bottom ash content	In improving tensile and flexural strength, chemical admixtures can be added to reduce water demand, resulting in greater tensile strength
Flexural strength	Arumugam et al. (2011)	Flexural strength of concrete decreased on substitution of sand with coal bottom ash	

2.2 Mix Design of SCC-BA

Self-compacting concrete incorporating bottom ash was designed to achieve the targeted strength of 40 MPa. A standard blended cement and 20 mm maximum coarse aggregate were considered for the study. A fine aggregate of 55% was used from the total aggregate weight. A superplasticizer (0.1–2%) was considered to improve the workability of the concrete mix. Different replacement percentages of CBA (10, 15, 20, 25, 30%) were used to evaluate the effect of CBA on concrete properties. Moreover, three water–cement ratios were chosen to produce flowable concrete (self-compacting concrete), as indicated in Table 2.2.

The local CBA was collected from Sultan Salahuddin Abdul Aziz Power Plant, Kapar, Selangor Darul Ehsan, as shown in Fig. 2.1a. The minimum size of CBA was found to be 0.075 mm, while the maximum size was 20 mm. Therefore, CBA was

Table 2.2 Mix proportions (kg/m^3)

w/c	Cement	Coarse aggregate	Fine aggregate	Water	Superplasticizer
0.35	557	715.5	874.5	194.95	1.11
0.40	518	715.5	874.5	207.20	1.04
0.45	485	715.5	874.5	218.25	1.12

Fig. 2.1 Raw coal bottom ash **a** collection, **b** sieved coal bottom ash

Table 2.3 Chemical composition of CBA

Compounds	Percentage composition (%)
Silicon dioxide, SiO_2	68.90
Aluminium dioxide, Al_2O_3	18.67
Iron oxide, Fe_2O_3	6.50
Calcium oxide, CaO	1.61
Potassium oxide, K_2O	1.52
Titanium dioxide, TiO_2	1.33
Magnesium oxide, MgO	0.53
Sodium oxide, Na_2O	0.24
Carbon dioxide, CO_2	0.10
Manganese oxide, MnO	–
Loss on ignition	1.72

graded by size passing through a sieve (5 mm), as shown in Fig. 2.1b. Moreover, the grading of CBA was important to avoid gap-graded in the mix.

Based on the particle dimensions, it was found that the CBA could be equally classified into two parts as 50% of CBA has a dimension less than 250 μm. This fact is considered an advantage such that CBA would fill the micropores present in the concrete matrix.

Table 2.3 presents the chemical composition of CBA. Kapar Energy Ventures supplied the CBA with a specific gravity of 1.90.

2.3 Properties of Fresh SCC-BA Subjected to Tap Water

2.3.1 Flowability

The slump test was conducted based on BS EN 12350-8. Both T500 and slump flow were considered to evaluate the flow rate and flowability without hindrances. From Fig. 2.2a, the slump flow for all SCC-BA mixture was found to be in the range 550–750 mm. The time corresponding to T500 slump flow was found to be less than 5 s for all SCC-BA mix, and the diameter reached up to 500 mm (Fig. 2.2b).

It was also noticed that the slump flow of the mixture decreased with an increase in coal bottom ash. Also, it was observed that the slump flow of SCC-10BA was slightly lower compared to that of the control mixture (0%). In contrast, there was a drop in the slump flow when the value of coal bottom ash replacement was greater than 10%. For the water–cement ratio of 0.35, the series of mixes 0-CBA, 10-CBA, 15-CBA, 20-CBA, 25-CBA, and 30-CBA demonstrated an average slump flow of 700 mm, 670 mm, 650 mm, 615 mm, 588 mm, and 545 mm, respectively. For w/c of 0.40, the average slump flows for the same series were 720 mm, 710 mm, 688 mm,

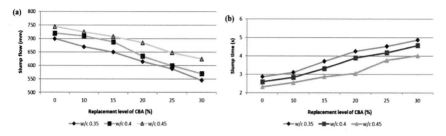

Fig. 2.2 Slump test **a** slump flow, **b** slump spread time (T500)

635 mm, 600 mm, and 570 mm, respectively. Meanwhile, the corresponding slump flow for a water–cement ratio of 0.45 for mixes of 0-CBA, 10-CBA, 15-CBA, 20-CBA, 25-CBA, and 30-CBA was 745 mm, 725 mm, 708 mm, 685 mm, 648 mm, and 625 mm, respectively. It was also noticed that increasing the replacement rate from 10 to 30% led to a decrease in the slump flow measurement by 4.29 and 22.14%. Similarly, the slump flow measurement dropped for specimens of 0.35 w/c by 1.39% and 20.83%. The extent of decrease was 2.68 and 16.11% for specimens of 0.40 water–cement ratio.

Therefore, it could be inferred that the presence of bottom ash greatly affected the concrete's workability due to the aggregates friction the physical morphology of coal bottom ash. In particular, the bottom ash has a rough texture and irregular shape. Also, CBA particles have a higher water absorption rate. This is because of the porosity of the CBA, wherein it absorbs the water present in the cement paste resulting in lower workability.

The T500 test was conducted to assess the rate of the concrete flow. For w/c of 0.35, the series of mixes 0-CBA, 10-CBA, 15-CBA, 20-CBA, 25-CBA, and 30-CBA witnessed an average slump spread time of 2.87 s, 3.1 s, 3.71 s, 4.26 s, 4.53 s, and 4.87 s, respectively. For a water–cement ratio of 0.40, the average slump spread times for the following series were 2.59 s, 2.84 s, 3.31 s, 3.90 s, 4.18 s, and 4.57 s, respectively. Meanwhile, the corresponding slump spread times for 0.45 water–cement ratio mixes of 0-CBA, 10-CBA, 15-CBA, 20-CBA, 25-CBA, and 30-CBA were 2.32 s, 2.55 s, 2.87 s, 3.05 s, 3.77 s, and 4.02 s, respectively. Apparently, the lower viscosity concrete will exhibit a quicker flow and eventually stop spreading, while higher viscosity concrete will spread forward over an extended time.

Figure 4.2 shows the influence of bottom ash on T500. It was evident that the T500 value decreased with the increase in the bottom ash content. A higher T500 value was observed due to the interparticle reaction between the aggregates. Clearly, the slump flow and slump spread time have shown that using the coal bottom ash in self-compacting concrete allowed replacement up to 30% while conforming to EFNARC standards for the flowability factor of self-compacting concrete. Hence, it can be concluded that the mixture with a higher water–cement ratio demonstrated a high slump flow value compared to the lower water–cement ratio. Besides, the presence of CBA in the concrete has negatively affected its workability due to the reaction between the aggregate particles.

2.3.2 Passing Ability

The flowability of the mixture is governed by the volume of the binder and the shape and size of the aggregate. Similarly, the passing ability results are also related to the aggregate accumulation in the narrow opening, which did not permit the aggregates particle to flow without hindrance. Therefore, the passing ability results would provide insight into the flowability of concrete mixture and be considered a necessary test representing the actual in situ condition of self-compacting concrete. Therefore, the passing ability of the SCC-BA was obtained and evaluated using the L-box test. It is acknowledged that the maximum aggregate size for this mixture was 16 mm to prevent any hindrance in the apparatus due to the blocking effect.

Figure 2.3 shows the passing ability of the fresh SCC incorporating different content of CBA. Based on the result, it was found that the passing ability ratios for all SCC-BA mixture were in the range of 0.6–1.0 depending on the amount of bottom ash used in which the value of passing ability of SCC-BA decreased with the increase of CBA replacement.

For w/c of 0.35, the mixtures of 0%, 10%, 15%, 20%, 25%, and 30% obtained from L-box ratio were 0.97, 0.92, 0.92, 0.88, 0.87 and 0.76, respectively. For water–cement ratio of 0.40, the L-box ratios for the same series were 0.94, 0.89, 0.86, 0.80, 0.80, and 0.69, respectively. Meanwhile, the corresponding L-box ratios for 0.45 water–cement ratio mixes of 0%, 10%, 15%, 20%, 25%, and 30% were 0.90, 0.81, 0.80, 0.77, 0.70, and 0.66, respectively.

The results showed that L-box passing ability of 0%, 10%, and 15% replacement level of CBA for all w/c met the recommendations stated by EFNARC. L-box ratios of 20% and 25% mixtures of coal bottom ash for 0.35 and 0.40 water–cement ratio followed the EFNARC recommendation. The mixture of 0.45 water–cement ratio obtained L-box ratio was below than 0.8. The mixture of 30% replacement of coal

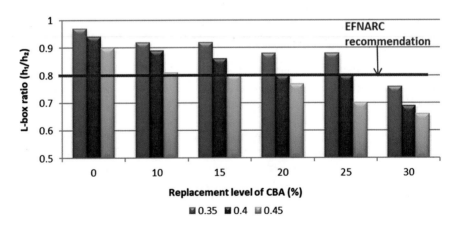

Fig. 2.3 L-box passing ability

bottom ash for all water–cement ratios was not in line with the standards. The aggregate blocking was detected in mixtures of 30%, 25%, and 20% replacement levels of CBA, resulting in a lower passing ability ratio. In the same context, it was found that the mixture of SCC-25BA, SCC-20BA, SCC-15BA, SCC-10, and SCC-0BA exhibited a good result in terms of passing ability test compared to SCC-30BA. In other words, the L-box ratios for the mixture of SCC-30BA were considered unsatisfactory.

2.3.3 Sieve Stability Test

The sieve stability test was conducted to assess the efficiency of the SCC-BA mixture to resist segregation. A sieve with a 5 mm square aperture was used to obtain the passing of the SCC-BA mixture. It is well known that a good ratio should fall within 5–15% of the sample's weight. If the mixture has a ratio lower than 5%, it indicates an excessive resistance that is likely to affect the surface finish. Meanwhile, ratios higher than 15% show that the mixtures have more possibility to be segregated. Figure 2.4 shows the segregation ratio index for SCC-BA mixtures.

For w/c of 0.35, the mixtures of 0%, 10%, 15%, 20%, 25%, and 30% demonstrated segregation ratios of 9.61%, 7.91%, 6.61%, 5.24% 5.20%, and 6.20%, respectively. For w/c of 0.40, the segregation ratios for the following series were 9.85%, 8.15%, 7.83%, 6.06%, 0.94, 0.89, 0.86, 0.80, 0.80, and 0.69, respectively. Meanwhile, the corresponding L-box ratios for 0.45 water–cement ratio mixes of 0%, 10%, 15%, 20%, 25%, and 30% were 0.90, 0.81, 0.80, 0.77, 0.70, and 0.66, respectively. It has been observed that the value of the segregation ratio decreased with the increase in CBA content.

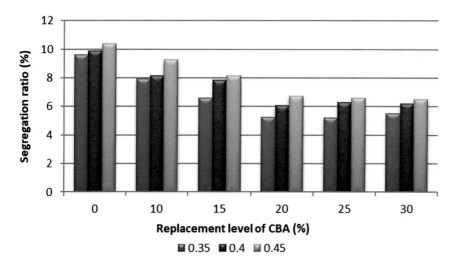

Fig. 2.4 Sieve stability test results

The mixture of 0% replacement level had the higher resistance index, which showed a better segregation resistance, while the lowest segregation ratio was achieved at 30% replacement of coal bottom ash. However, the value of the 10% replacement level was slightly lower than 0%. The increase in the replacement level of coal bottom ash has resulted in a significant decrease in the segregation index value. Thus, it is possible to claim that the aggregate would not settle down with the increase in coal bottom ash content.

In terms of water–cement ratios, the result showed a similar pattern for all mixtures. The segregation ratio of 0.35 water–cement ratio on a mixture of 20, 25, and 30% replacement was below 5%. Meanwhile, the segregation ratio for mixture with w/c of 0.40 and 0.45 ranged from 6 to 10%.

2.4 Properties of Hardened CBA Self-Compacting Concrete

2.4.1 Unit Weight

Table 2.4 presents the recorded unit weight results for three different water/cement ratios of self-compacting concrete. It can be observed that the minimum concrete unit weight was 2125 g/m^3, while the maximum value reached up to 2436 g/m^3. The average concrete density with coal bottom ash for the aggregate was 2330 g/m^3 at 28 days which showed a decrease of 3.1% compared to a control concrete density of 2404 g/m^3. The results revealed that the control mixture acquired the highest density while the SCC-30BA mixture showed the lowest density. The coal bottom ash inclusion in the concrete mixture reduced the unit weight density irrespective of the ages of curing. The increase in coal bottom ash aligned with the decrease in the density of concrete specimens.

Another perspective of the results revealed that the incorporation of CBA in the mixture affected concrete workability due to the morphology of the CBA. CBA, being a porous material, enables water to be absorbed and, as a result, reduces workability. With an increase in the coal bottom ash, the workability was reduced. It could also be seen that the decrease in the concrete density increased with an increase in the CBA replacement level.

On the other hand, higher concrete density was achieved for a lower water–cement ratio. Based on the results, for w/c of 0.35, the reduction in concrete density concrete was found to be 2.86% for the ageing of 28 days. Besides, for 0.45 and 0.4 w/c, the reduction in concrete density was 3.2% and 3.01%, respectively. In general, the unit weight of SCC-BA was relatively low (Fig. 2.5) and had a porous structure compared to natural sand. The inclusion of a higher amount of coal bottom ash leads to an increase in the mixing water, subsequently increasing the size of the concrete pores.

Table 2.4 Unit weight of specimen (g/cm^3)

Curing age	w/c	Replacement of coal bottom ash					
		0%	10%	15%	20%	25%	30%
7 days	0.35	2394	2336	2319	2301	2295	2286
	0.40	2363	2241	2220	2201	2190	2169
	0.45	2317	2202	2195	2173	2146	2125
14 days	0.35	2414	2373	2352	2322	2314	2318
	0.40	2388	2355	2331	2317	2300	2287
	0.45	2372	2332	2300	2284	2271	2249
28 days	0.35	2426	2389	2371	2357	2339	2327
	0.40	2403	2361	2346	2325	2312	2295
	0.45	2383	2348	2319	2302	2289	2276
60 days	0.35	2431	2393	2379	2364	2349	2352
	0.40	2412	2370	2355	2332	2327	2302
	0.45	2403	2352	2326	2310	2294	2284
90 days	0.35	2435	2401	2383	2369	2353	2358
	0.40	2422	2383	2359	2337	2331	2309
	0.45	2408	2355	2329	2312	2300	2292
180 days	0.35	2436	2402	2385	2372	2355	2360
	0.40	2423	2385	2362	2341	2333	2310
	0.45	2407	2357	2331	2311	2302	2362

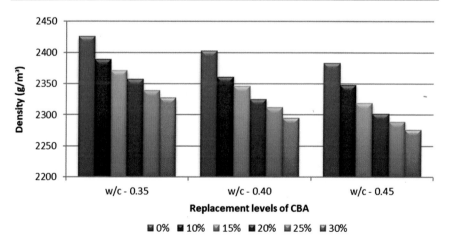

Fig. 2.5 Density of concrete at 28 days

2.4.2   Compressive Strength

The compressive strength of concrete is an essential indicator in civil engineering. It is considered a critical parameter that reflects the capacity and consistency of the material ingredients of a concrete mix. The compressive strength also has a direct relationship with the quality and properties of the mix. In the present section, the compressive strength was taken into account to evaluate the extent to which the CBA could affect the self-compacting concrete. Cubes (100 × 100 × 100 mm) were cast after mixing. The target cubes were then cured after 24 h in a water tank and tested for ageing of 28, 60, 90, and 180 days according to BS EN 12390-3:2009 and BS EN 12390 4:2000. The results of the compressive strength of SCC-BA subjected to tap water are presented in Table 2.5. A significant decrease in concrete strength was obtained in the addition of high content of CBA. For example, the decrease in strength for sample 20–30% coal bottom ash replacement could have resulted from the physical morphology of CBA. Specifically, CBA is a porous material that could negatively affect the concrete microstructure.

After 28 days, the concrete mixture of 0.35 w/c endured 35.2, 38.2, 39.5, 42.0, 58.0, and 57.3 MPa for 30%, 25%, 20%, 15%, 10%, and 0% replacement of bottom ash, respectively. The compressive strength of SCC-BA with 0.40 w/c was obtained

Table 2.5 Results of compressive strength test

Curing age	w/c	Replacement of coal bottom ash					
		0%	10%	15%	20%	25%	30%
7 days	0.35	42.7	43.3	38.6	35.1	33.9	31.2
	0.40	35.1	37	33.2	32.1	29.9	28.3
	0.45	28.4	29	25.2	23.8	22.3	21.8
14 days	0.35	53.3	56.7	42.3	38	36.8	35.2
	0.40	45.6	48	36.7	33.2	31.7	29.5
	0.45	39.1	43.4	32.9	31.3	28.6	25.6
28 days	0.35	57.3	58	42	39.5	38.2	35.2
	0.40	47.2	48.7	38	36.5	34.2	31.6
	0.45	41	44	34.4	32.9	31.2	30
60 days	0.35	65	67.2	45.1	41.7	40.8	39.2
	0.40	50.3	53.5	42.9	38.4	36.3	34.2
	0.45	42.7	47	37.2	34.7	32.9	31.6
90 days	0.35	68	69.8	50.4	47.2	45.6	44.6
	0.40	60.4	64.3	47.2	43	41.2	40.8
	0.45	51.2	54.3	44.5	42.1	40.4	41
180 days	0.35	70.3	71.4	53.7	51.2	49.4	46.7
	0.40	63.6	67.8	50.4	48.9	47.5	45.5
	0.45	52	55	46.4	44.6	42.1	41

at 31.6, 34.2, 36.5, 38.0, 48.7, and 47.2 MPa for 30%, 25%, 20%, 15%, 10%, and 0% replacement CBA, respectively. Meanwhile, the strength of SCC-BA with 0.45 w/c concrete was found to be at 30, 31.2, 32.9, 34.4, 44.0, and 41.0 MPa for 30%, 25%, 20%, 15%, 10%, and 0%, respectively. From the findings, it is evident that the compressive strength of the SCC-10BA mixture was slightly higher compared to other mixtures. The self-compacting concrete with 10% coal bottom ash replacement displayed the highest compressive stress endurance, which was at least 40 MPa at 28 days for all water–cement ratios. Also, the compressive strength of SCC-10BA specimens with water–cement ratios of 0.45, 0.40, and 0.35 was about 9.77%, 3.08%, and 3.37%, respectively, which were higher than that of the control specime for the ageing duration of 28 days, respectively.

Compressive strength results for ageing ages of 7, 14, 28, 60, 90, and 180 days are illustrated in Fig. 2.6a–c. At an ageing duration of 180 days, the compressive strength of SCC-BA with 0.35 w/c measured about 70.3, 71.4, 53.7, 51.2, 49.4 and 46.7 MPa for the mixtures of 0%, 10%, 15%, 20%, 25% and 30% replacement, respectively. For 0.40 and 0.45 w/c mixtures, the strength at 180 days was marginally lower compared to the strength of the 0.35 water–cement ratio mixture. The mixture of 10% coal bottom ash replacement demonstrated a higher compressive strength than the 0% mixture. This is attributed to the increase in the fine particle content and the pozzolanic reaction of bottom ash inside the concrete matrix. The cement past and interfacial transient zone contain numerous pores. These pores can be classified according to their size. Based on Zhang et al. (2018), three levels represented the pores according to their diameter. Gel pores represent the smallest pores with a diameter of less than 10 nm. Gel pores are ascribed to the hydration process, and they are important, as they play a great role in affecting porosity and durability.

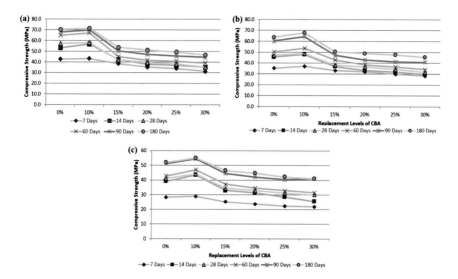

Fig. 2.6 Compressive strength for specimens of **a** 0.35 w/c, **b** 0.40 w/c, **c** 0.45 w/c

The second level comprises capillary pores, which usually range between 10 and 1000 nm in size. Capillary pores are defined as the original water space, which is not filled by the hydrated cement product. These pores can be an important factor deterring concrete strength. Finally, air voids are regarded as the third level, and they are greater than 1000 nm in size. The two main causes of air voids are inadequate compaction and deliberately entrained air. The positive enhancement of SCC-BA strength was due to the plugging of the concrete pores and the physical morphology of coal bottom ash. However, the value of SCC-BA strength enhancement was not significant. Such results confirmed that adding coal bottom ash to the concrete was to enhance the durability of concrete exposed to aggressive conditions, which has been discussed in detail in the next chapters.

It can be summarized that the optimum replacement of sand by CBA in self-compacting concrete was 10%. The reason behind the increase in the compressive strength of SCC-BA was the pore refinery effect by the pozzolanic reaction of CBA. In other words, the silica content in CBA particles plays a significant role in forming extra gel C-S-H, which is responsible for the increase in the strength. The compressive strength of SCC-BA notably decreased as a result of all these factors. Since the CBA is a porous material, it resulted in the reduction of the compressive strength. In terms of water–cement ratio, a higher compressive strength was recorded for a lower cement ratio.

2.4.3   Flexural Strength

Table 2.6 presents the results of the flexural strength of self-compacting concrete containing different amounts of CBA and water–cement ratios. The flexural strength was recorded at the age of 7, 14, 28, 60, 90, and 180 days. It was interesting to notice that the highest flexural strength was obtained for 10% replacement of fine aggregate by coal bottom ash compared to that of the control specimens.

With regard to the water–cement ratio, the enhancement of the flexural strength of SCC with 10% CBA and the water–cement ratio of 0.35 was 1.58% and 9.2% at 28 and 180 days, respectively. Moreover, the increment in flexural strength of SCC-10BA with 04 w/c was found to be 4.41% and 15.91% at 28 and 180 days, respectively. Meanwhile, at the age of 28 and 180 days, the flexural strength of SCC-10BA with a water–cement ratio of 0.45 was observed to have enhanced by 2.03% and 17.82% compared to that of the control mixture.

Conversely, as the replacement percentage of coal bottom ash increased for all water–cement ratios, a decrease in the flexural strength was recorded for the concrete. The result was attributed to the poor joining between aggregates inside the concrete matrix. The extent of the decrease in the flexural strength of SCC-BA depends on the replacement level of coal bottom ash.

2.4.4   Split Tensile Strength

Table 2.7 presents the split tensile strength results of self-compacting concrete incorporated with CBA. A partial replacement of CBA as fine aggregate in self-compacting concrete had enhanced the tensile strength for all curing age durations, in line with the results of compressive strength. The increment in tensile strength of SCC-BA was attributed to the decrease in the pores and microcracks inside the concrete mixture. Moreover, it was noticed that the improvement in the tensile strength is curing age-dependency, which was also in good agreement with the findings of Zainal Abidin et al. (2015).

On the other hand, it was found that the tensile strength reduced when the water–cement ratio increased. For a simpler analysis, the results of the split tensile strength of self-compacting concrete containing CBA are graphically plotted in Fig. 2.7a–c. It was observed that the highest tensile strength was recorded for the water–cement ratio of 0.35 compared to that of the other ratios. At curing age of 28 days and water–cement ratio of 0.35, the split tensile strength of SCC containing 30%, 25%, 20%, 15%, 10%, and 0% of coal bottom ash as a fine aggregate recorded 3.03, 3.18, 3.50, 3.86, 4.50, and 4.25 MPa, respectively. For water–cement ratio of 0.4, the split tensile strength of SCC incorporating 30%, 25%, 20%, 15%, 10%, and 0% of CBA

Table 2.6   Results of flexural strength

Curing age	w/c	Replacement of coal bottom ash					
		0%	10%	15%	20%	25%	30%
7 days	0.35	6.03	6.3	5.09	4.51	4.38	4.17
	0.40	5.47	5.61	4.86	4.1	3.92	3.78
	0.45	4.86	4.92	4.17	3.89	3.73	3.59
14 days	0.35	6.97	7.06	5.89	5.7	5.58	5.19
	0.40	5.65	5.81	5.01	4.56	4.38	4.25
	0.45	5.3	5.5	4.52	4.29	4.11	3.9
28 days	0.35	7.49	7.62	6.7	6.21	6.04	5.87
	0.40	6.5	6.8	5.82	5.31	5.02	4.89
	0.45	6.27	6.4	5.4	5	4.71	4.66
60 days	0.35	7.56	7.87	6.92	6.34	6.29	6.02
	0.40	7.31	7.42	6.05	5.58	5.33	5.13
	0.45	6.87	6.99	5.75	5.34	5.01	4.96
90 days	0.35	7.72	8.02	7.01	6.62	6.37	6.26
	0.40	7.48	7.55	6.84	6.12	5.74	5.34
	0.45	7.02	7.11	6.32	5.53	5.38	5.18
180 days	0.35	7.84	8.26	7.34	7.01	6.89	6.45
	0.40	7.62	7.73	6.92	6.45	6.12	5.84
	0.45	7.43	7.63	6.7	6.11	5.5	5.26

Table 2.7 Results of split tensile strength

Curing age	w/c	Replacement of coal bottom ash					
		0%	10%	15%	20%	25%	30%
7 days	0.35	3.05	3.15	2.7	2.57	2.41	2.35
	0.40	2.96	3.05	2.47	2.4	2.35	2.29
	0.45	2.3	2.45	2.05	1.87	1.74	1.69
14 days	0.35	3.61	3.7	3.11	2.87	2.73	2.6
	0.40	3.29	3.42	2.89	2.62	2.54	2.48
	0.45	3.01	3.15	2.44	2.31	2.15	2.02
28 days	0.35	4.25	4.5	3.86	3.5	3.18	3.03
	0.40	3.81	3.98	3.34	2.93	2.74	2.7
	0.45	3.43	3.56	3.03	2.78	2.53	2.41
60 days	0.35	4.64	4.71	4.1	3.7	3.31	3.12
	0.40	4.03	4.24	3.78	3.3	3.12	3.09
	0.45	3.83	3.97	3.22	2.85	2.77	2.69
90 days	0.35	4.73	4.92	4.26	3.74	3.58	3.31
	0.40	4.32	4.63	4.28	3.41	3.23	3.12
	0.45	3.94	4.21	3.3	3.09	2.98	2.73
180 days	0.35	4.84	5.1	4.6	3.99	3.67	3.55
	0.40	4.45	4.87	4.34	3.75	3.41	3.33
	0.45	3.98	4.31	3.65	3.15	3.08	3.05

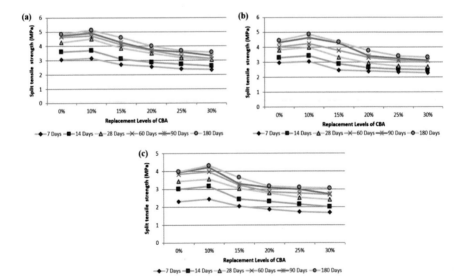

Fig. 2.7 Split tensile strength for specimens of **a** 0.35 water–cement ratio, **b** 0.40 water–cement ratio, **c** 0.45 water–cement ratio

was measured at 2.70, 2.74, 2.93. 3.34, 3.98, and 3.81 MPa, respectively, while for the radio of 0.45, the split tensile strength of SCC containing 30%, 25%, 20%, 15%, 10%, and 0% of CBA was recorded at 2.41, 2.53, 2.78, 3.03, 3.56, and 3.43 MPa, respectively.

The optimum tensile strength result was recorded for 10% of bottom ash replacement in which the tensile strength of SCC-10BA exhibited an increase compared to concrete specimens without coal bottom ash. In contrast, the tensile strength of SCC incorporating 15, 20, 25, and 30% of CBA gradually decreased because of an increase in replacement level of CBA as the fine aggregates had formed porous concrete structure. With more pores scattered around the CBA aggregate surface, more reduction in tensile strength was observed.

The substitution of fine aggregates with coal bottom ash affected the split tensile strength differently. For example, the bonding between cement paste and aggregate affected the strength. The porous-rough surface aggregate, which usually has softer, spongy, and mineralogically dissimilar elements, was found to form a robust aggregate-cement paste bond compared to normal aggregate.

2.4.5 Water Absorption and Permeable Pore Space

Figure 2.8a–c shows the performance of self-compacting concrete containing different amount of CBA using permeable pore space and water absorption tests. The replacement of sand by CBA in self-compacting concrete witnessed a significant change for all water–cement ratios. The percentage of the water absorption and permeable pore space of concrete containing 10% of CBA had improved compared to that of the control concrete. Moreover, the water absorption and permeable pore spaces of concrete increased with an increase in CBA content, specifically after 28 days. In particular, the water absorption and permeable pore space showed an increment with higher coal bottom ash content from 15 to 30%.

For water–cement ratios of 0.35, 0.40, and 0.45, the permeable pore space in concrete increased from 10.35 to 10.74%, 10.56 to 10.86%, and 10.60 to 11.04% after 28 days, respectively. Meanwhile, water absorption of bottom ash concrete mixtures

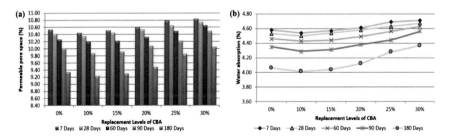

Fig. 2.8 Specimens of 0.35 water–cement ratio **a** permeable pore space, **b** water absorption

increased from 4.50 to 4.67%, 4.59 to 4.72%, and 4.61 to 4.75%. The findings were similar to that of the study conducted by Singh and Siddique (2015). The study revealed that the more the content of coal bottom ash, higher the enhancement of connectivity between the capillaries. In particular, the connectivity between the capillaries of concrete incorporating coal bottom ash was significantly improved at the ageing of 28 days. After that, the permeable pore space and water absorption reduced with the progress of curing age. In other words, the reduction in permeable pore space in self-compacting concrete containing CBA was curing age-dependence. It is also interesting to note that the pozzolanic activity of the CBS was responsible for the decrease of permeable pore space in coal bottom ash self-compacting concrete mixtures. Also, the results showed a good correlation between water absorption and permeable pore space.

2.4.6 Compressive Strength Relationship

2.4.6.1 Relationship Between Compressive Strength and Density

Figure 2.9 shows the relationship between density and compressive strength. To evaluate the strength of the relationships, the determination coefficient (R^2) was employed as a statistic estimator. R^2 can provide insight into the degree of fitting between the density and the compressive strength. Its value is generally in the range of 0–1. Moreover, the best fitting of density and the compressive strength results could

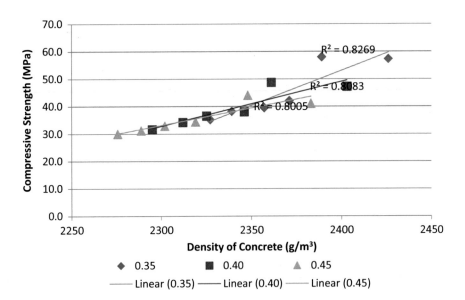

Fig. 2.9 Relationship between compressive strength and concrete density

be obtained by increasing the value of the determination coefficient. According to the logical hypothesis, higher accuracy of fit between density and compressive strength was obtained when the correlation coefficients were greater than 0.8. The values of the coefficient of determination for water–cement with the ratios of 0.35, 0.40, and 0.45 were 0.83, 0.81, and 0.80 are estimated. As such, the relationship of density and compressive strength can be considered close results.

On the other hand, it has been observed that the value of concrete unit weight ranges between 2276 and 2426 kg/m³. Interestingly, the increase in CBA content corresponds with the decreasing density of the concrete specimens.

Furthermore, the workability was dependent on the amount of coal bottom ash in the mixture, in which the workability decreased with an increase in the CBA content. The high variation in the water–cement ratio would affect the concrete density, and compressive strength, wherein a lower water–cement ratio resulted in a higher density and compressive strength. This is because the pore refinement activity has reduced the concrete's porosity by the pozzolanic reaction of CBA due to the considerable reduction in concrete density; thus, it exhibits lower compressive strength.

2.4.6.2 Relationship Between Split Tensile and Compressive Strength

Figure 2.10 shows the relationship between the split tensile strength and the compressive strength. The results have indicated that the compressive strength of specimens increased linearly with the increase in tensile strength. The decrease in the compressive strength from 57.3 to 35.2 MPa indicates that the higher content of CBA in SCC

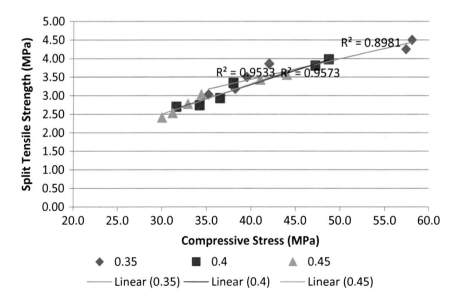

Fig. 2.10 Relationship between split tensile strength and compressive strength

would result in a lower strength. By increasing the content of CBA, tensile strength decreased compared to that of the control mixture.

The results also showed that the R^2 values for water–cement ratios of 0.45, 0.40, and 0.35 are 0.95, 0.96, and 0.90, respectively. As the value of coefficients of determination R^2 was near 1, the relationship between the compressive strength and splitting tensile strength was considered reliable and accepted for all water–cement ratios. In addition, according to the statistical indicators, the relationship of split tensile strength and compressive strength was also considered close results as the value of R^2 was greater than 0.7.

2.4.6.3   Relationship Between Flexural and Compressive

As discussed earlier, it was found that the compressive strength of SCC was strongly linked to the amount of coal bottom ash, while it decreased with an increase in CBA content. The decrease in compressive strength was due to the high content of CBA in self-compacting concrete. Similarly, the increment in sand replacement by coal bottom ash showed a reduction in flexural strength than the control mixture. Based on Fig. 2.11, the experimental result has also indicated that the compressive strength of SCC-BA increased linearly with the increase in tensile strength. Also, the reliability of the linear relationship between the flexural strength and compressive strength was imparted by the coefficient of determination R^2. It was found that R^2 values for water–cement ratios of 0.45, 0.40, and 0.35 were 0.97, 0.97, and 0.97, respectively, which indicated that the compressive strength fit relatively well to the split tensile strength. It could be inferred that the relationships between flexural strength and

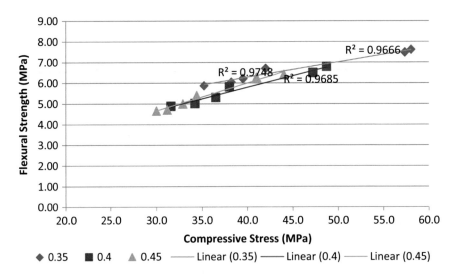

Fig. 2.11 Relationship between flexural strength and compressive strength

compressive strength in terms of determination coefficients were acceptable for all water–cement ratios.

2.4.6.4 Relationship Between Water Absorption and Compressive

The good strength of concrete can be achieved with low water absorption. Besides, concrete should exhibit good performance in terms of workability, strength, and durability. The correlation should also be strong between compressive strength, water absorption, and porosity. In general, self-compacting concrete has a dense structure of hydrated cement paste which contains a discontinuous capillary pore system. Thus, the concrete poses great resistance to external attack.

Figure 2.12 shows a good relationship demonstrated by the compressive strength with water absorption. The results indicated that the compressive strength of SCC-BA improved linearly with the increase in water absorption. The results revealed that the coefficient of determination R^2 values for water–cement ratios of 0.45, 0.40, and 0.35 was 0.7593, 0.9573, and 0.7038, respectively. The decreased compressive strength was attributed to the high content of coal bottom ash in SCC, leading to higher water absorption. Moreover, the increased replacement level of CBA as fine aggregate showed an increase in water absorption compared to the control. The correlation between compressive strength and water absorption was found to be reliable and was recognized for all water–cement ratios with the value of R^2 close to 1.

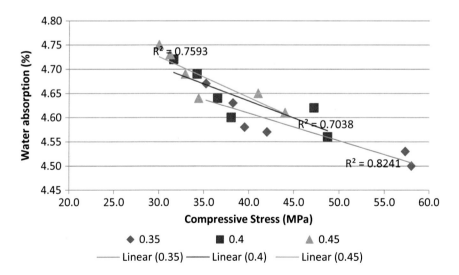

Fig. 2.12 Relationship between water absorption and compressive strength

2.5 X-Ray Diffraction

X-ray powder diffraction (XRD) is a non-destructive tool utilized for the phase identification of any crystalline material by recognizing the arrangement of atoms. Briefly, the phase identification is carried out by directing the generated X-rays to the target powder specimen that contains many crystals located at random angles of atomic planes. Then, the interaction of crystalline atoms with the generated X-rays produces diffracted X-rays in different directions at the same wavelength, resulting in the formation of interference patterns. The created interference patterns can be classified as destructive or constructive, depending on the distance between the atom layers, angle of diffraction, and chemical composition. Also, Bragg's law is used to obtain the inference patterns and the diffraction peak. Finally, the target specimen can be identified by comparing the intensity and position of the diffraction peaks with standard reference patterns in the International Centre for Diffraction Data.

The concrete intermediate zone is different from the cement paste. The small diffraction peaks of calcium silicate hydrate were observed to be crowded by the diffraction peaks in the oxide mineral called portlandite or $Ca(OH)_2$. In the present study, the XRD technique was conducted for examination of the evolved phases in the concrete after 28 and 180 days. The diffraction angle 2θ of XRD analysis ranged between $10°$ and $90°$ in steps of $2\theta = 0.017°$. The XRD analysis of powder cement paste of concrete incorporating 0, 10, 15, 20, 25, and 30% coal bottom was conducted.

The hydrated cement products such as calcium silicate hydrate, ettringite, quartz, calcium hydroxide, calcium silicate, and calcium alumina silicate hydrate were identified in the diffractograms analysis. The presence of ettringite was detected in all concrete samples which could be probably due to the internal sulphate attack ions. As observed from diffractograms, tricalcium silicate (alite) and dicalcium silicate (belite) and calcium silicate hydrate were also identified. The diffraction peaks of hydration phases in concrete paste were affected by silicon oxide, which occurred in the aggregates bonding.

The amount of calcium silicate was higher in the control sample than the sample containing coal bottom ash. The total amount of ettringite in samples had not changed in the partial replacement of coal bottom ash to natural sand. According to Singh and Siddique (2014), coal bottom ash has an important role in the formation of the portlandite phase. $Ca(OH)_2$ forms a mixture with the CSH gel that reduces the pore volume by converting the liquid water into solid form. The cement paste was exposed to the water, and then $Ca(OH)_2$ dissolved, resulting in a porous material. The peak intensity of portlandite in coal bottom ash mixtures 0%, 10%, 15%, 20%, 25% and 30% was 3120, 4011, 3935, 3063, 3264, and 2198 counts, respectively. For example, the total amount of portlandite increased as the coal bottom ash content increased in the concrete at 28 days of curing age (Fig. 2.13a). Conversely, the total amount of portlandite decreased at ageing period of 180 days (Fig. 2.13b). Thus, portlandite reduction in the concrete samples is marked by the activity of coal bottom ash pozzolanic. On comparing the coal bottom ash replacement levels, the amount of portlandite decreased with higher content of coal bottom ash. This inverse

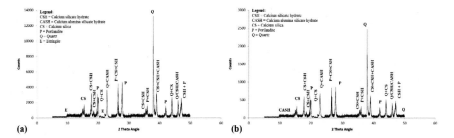

Fig. 2.13 XRD analysis of concrete with 10% coal bottom ash at **a** 28 days and **b** 180 days

relationship has often been found possibly due to the available content, which does not correspond to the total portlandite content since the variable proposition is embedded by the hydrates as a result of non-participative nature in the pozzolanic reaction.

2.6   Scanning Electron Microscopy

The concrete microstructure has a significant impact on their mechanical, physical properties, and durability. The quantity and distribution of pores present in the cement paste, aggregate, and interfacial transition zone are closely related to the concrete microstructure. Therefore, the scanning electron microscopy (SEM) was employed to examine and observe the concrete microstructure, while EDX analysis was used to identify the chemical products that were present inside the concrete matrix. In other words, an SEM analysis is essential to evaluate the concrete microstructure. This is because a poor concrete microstructure not only affects the concrete strength but also provides an easy path for aggressive material to attack the concrete and thus affects the lifespan of concrete structures. SEM equipped with EDX was used to assess the microstructure of self-compacting concrete incorporating bottom ash.

Self-compacting concrete microstructure features are affected by the type of cement, hydration period, and the mineral admixtures used. In this study, a ruptured portion of concrete produced from the compressive strength tests was attached to the scanning electron microscopy stub, and the images were taken by secondary electron image technique.

Indeed, the calcium silica hydrate (CSH) is a major phase present in all concrete samples. Based on the obtained results, there was not much dissimilarities present in the concrete associated with the difference in water–cement ratios. Besides, a significant contradiction was found in the concrete sample incorporated with different levels of coal bottom ash replacement. Figure 2.14a shows an indication of thick, dense, and continuous CSH gel, ettringite needles, and crystals of portlandite bonds in the control concrete. This characteristic feature was also observed in the concrete with 10% coal bottom ash shown in Fig. 2.14b.

Fig. 2.14   SEM images of concrete incorporating 10% of CBA at 28 days

Moreover, with an increase in the sand replacement content (20–30%), a dense CSH structure was observed. However, the CSH gel structure was considerably less uniform compared to CSH gel in the control concrete. The characterization of non-uniform microstructure is due to the small-sized pores (5–50 μm) and portlandite crystals present in coal bottom ash self-compacting concrete mixtures. The progress of ettringite in the voids was exposed in both control mix and self-compacting concrete containing CBA. The slight decrease in compressive strength of coal bottom ash self-compacting concrete at 28 days of curing age could be due to the variations in the microstructure samples from that of control concrete because of slow pozzolanic activity of coal bottom ash.

2.7   Conclusion

Self-compacting concrete is crucial to the achievement of sustainable concrete structures. The replacement of the fine aggregate by the coal base has shown the potential of the waste product to be used as construction materials. The details of the findings were presented and discussed in terms of raw properties, hardened properties, phase analysis (XRD), and microstructural and chemical analyses (SEM–EDX). The results showed that the L-box ratio, slump flow, and sieve segregation resistance of self-compacting concrete incorporating coal bottom ash have decreased; meanwhile, the slump spread has increased with the increase in coal bottom ash. The behaviour of compressive strength, flexural strength, and tensile strength has shown a similar pattern wherein they decreased with an increase in coal bottom ash replacement. This result was attributed to the physical morphology of coal bottom ash wherein it is regarded as porous material. On the other hand, it was found that only the 10% replacement mixture showed a higher result compared to the control. In terms of water–cement ratios, it was found that the low water–cement ratio showed higher strength and durability. However, a lower water–cement ratio may not mix thoroughly and not flow well enough to be sited in self-compacting concrete. The water–cement ratio of 0.40 is therefore considered to be the suitable ratio in order to ensure a fully hydrated reaction process.

References

Abubakar, A. U., and K. S. Baharudin. 2012. Properties of concrete using tanjung bin power plant coal bottom ash and fly ash. *International Journal of Sustainable Construction Engineering and Technology* 3: 56–69.

Aggarwal, Y., and R. Siddique. 2014. Microstructure and properties of concrete using bottom ash and waste foundry sand as partial replacement of fine aggregates. *Construction and Building Materials* 54: 210–223.

Arumugam, K., R. Ilangovan, and D. J. Manohar. 2011. A study on characterization and use of pond ash as fine aggregate in concrete. *International Journal of Civil & Structural Engineering* 2: 466–474.

Ibrahim, M. W., A. F. Hamzah, N. Jamaluddin, P. Ramadhansyah, and A. Fadzil. 2015. Split tensile strength on self-compacting concrete containing coal bottom ash. *Procedia-Social and Behavioral Sciences* 195: 2280–2289.

Singh, M., and R. Siddique. 2013. Effect of coal bottom ash as partial replacement of sand on properties of concrete. *Resources, Conservation and Recycling* 72: 20–32.

Singh, M., and R. Siddique. 2014. Strength properties and micro-structural properties of concrete containing coal bottom ash as partial replacement of fine aggregate. *Construction and Building Materials* 50: 246–256.

Singh, M., and R. Siddique. 2015. Properties of concrete containing high volumes of coal bottom ash as fine aggregate. *Journal of Cleaner Production* 91: 269–278.

Topçu, I. B., M. U. Toprak, and T. Uygunoğlu. 2014. Durability and microstructure characteristics of alkali activated coal bottom ash geopolymer cement. *Journal of Cleaner Production,* 81: 211–217.

Zainal Abidin, N. E., M. H. Wan Ibrahim, N. Jamaluddin, K. Kamaruddin, and A. F. Hamzah. 2015. The strength behavior of self-compacting concrete incorporating bottom ash as partial replacement to fine aggregate. *Applied Mechanics and Materials*, 2015. trans tech publ, 916–922.

Zhang, J., F. Bian, Y. Zhang, Z. Fang, C. Fu, and J. Guo. 2018. Effect of pore structures on gas permeability and chloride diffusivity of concrete. *Construction and Building Materials* 163: 402–413.

Chapter 3
CBA Self-compacting Concrete Exposed to Chloride and Sulphate

Abstract The penetration of harmful chemical ions such as sulphate and chloride ions into the concrete matrix is one of the leading causes of steel reinforcement corrosion. This has considerably affected both the compressive strength and service life of the concrete structure. As a result, cement-based concrete structures exposed to harsh conditions should be appropriately designed to achieve high strength and meet durability requirements. One of the sustainable strategies for prolonging a concrete lifespan is the use of coal bottom ash in self-compacting concrete. A brief background of both the sulphate attack and the chloride attack mechanism is highlighted in this chapter. The effects of different CBA replacement levels as fine aggregates in SCC exposed to sodium sulphate (Na_2SO_4) and sodium chloride (NaCl) solution are discussed in detail.

3.1 Mechanism of Sulphate Attack

The sulphate attack is one of the most common salt attacks that cause damage to concrete and therefore affect concrete durability. Sulphate attacks, as is well known, are related to the formation of ettringites and gypsum. Specifically, the aluminate present in the hardened concrete matrix reacts with the sulphate ions present in the surroundings to form ettringite and gypsum. These two products generate internal stress due to the volumetric change inside the concrete pores resulting in cracking that appears on the concrete surface. Certain factors such as permeability, the presence of mineral admixture, the type and content of cement, and the concentration of sulphate solution also play a major role in limiting the impact of sulphate attacks on the concrete matrix.

© The Author(s), under exclusive license to Springer Nature Singapore Pte Ltd. 2021   33
M. H. Bin Wan Ibrahim et al., *Properties of Self-Compacting Concrete*
with Coal Bottom Ash Under Aggressive Environments, SpringerBriefs in Applied
Sciences and Technology, https://doi.org/10.1007/978-981-16-2395-0_3

3.2 Mechanism of Chloride Attack

The penetration of chloride ions into the concrete matrix is also considered to be a major reason causing the reinforcement's corrosion. The effect of chloride ions on hardened concrete can have adversely effects in many ways. Generally, the formation of chloroaluminate or Friedel salt is the main adverse effect that leads to an increase in concrete permeability.

Regarding the mechanism of reinforcement corrosion, carbon dioxide, chloride ion, oxygen, and water from the atmosphere tend to penetrate and reach the steel reinforcement inside the concrete matrix. Then, the alkaline pore solution begins to reduce, leading to the destruction of the passivity layer. Finally, the steel reinforcement corrodes and generates a volumetric change that leads to a concrete spalling. Such negative results lead to a deterioration of the concrete structure. Therefore, concrete exposed to aggressive conditions, such as chloride ions, must be more durable to withstand degradation modes.

3.3 Mix Design of CBA Self-compacting Concrete

3.3.1 Material Proportions

A standard blended cement and 20 mm maximum coarse aggregate were used in all concrete mixture to achieve targeted strength of 40 MPa. The ratio of fine aggregate is 55% of the total aggregate weight. A superplasticizer was also added in order to improve the concrete workability. The proportions of the involved material were designed based on the method used by Jawahar et al. (2012). Three water–cement ratios (0.45, 0.4, 0.35) were considered. For cyclic wetting and drying, a water–cement ratio of 0.40 was used to study the exposure of self-compacting concrete to an aggressive environment.

3.3.2 Curing Set-up

All the concrete samples were cured in water maintained at room temperature to achieve a strength of 40 MPa at the age of 28 days. After that, the specimens were transferred to specific exposure conditions. The samples were exposed separately to distinct wetting and drying cycles.

It is important to note that the concrete specimens exposed to wetting cycles were sited in a square of 10 L plastic containers filled with 5% NaCl solution, 5% Na_2SO_4 solution, or seawater. The selection of the chemical concentration can cause severe damage for tested mixture. It is performed this way to accelerate the results that correspond to the field performance of concrete under harsh conditions within a

Table 3.1 Chemical composition of sodium sulphate

Compounds	Percentage composition (%)
Assay	99.47
Insoluble	0.02
Loss on ignition	0.5
Chloride, Cl	0.005
Nitrate, NO_3	0.003
Phosphate, PO	0.002

reasonable testing period in the laboratory. All cyclic tests were conducted in identical 10 L plastic containers. It should also be noted that both drying and wetting cyclic tests were conducted at a laboratory temperature of (25–31 °C). For cyclic exposure, the wetting duration of 15 h was taken into account, while the drying time was kept for 9 h. In particular, the SCC-BA samples were first wetted for 15 h in the target solution. After that, the SCC-BA samples were shifted to another enclosed chamber for the purpose of cyclic drying (9 h). Then, SCC-BA samples were returned to the wetting exposure. This process was carried out and repeated until the exposure duration of cyclic condition was completed. Five stages of ageing (7, 28, 60, 90, and 180 days) were considered for testing the SCC-BA samples.

3.3.3 Sodium Sulphate

According to ACI 318-08, 5% of sodium sulphate represents severe sulphate exposure. The sodium sulphate (Na_2SO_4), a white crystalline solid, also known as an anhydrous mineral thenardite, is a product of sulphuric acid. In addition, sodium sulphate is a white crystalline solid. In the laboratory, the Na_2SO_4 was diluted with distilled water (5% by weight of volume). The chemical analyses of sodium sulphate are tabulated in Table 3.1. At each cycle of concrete exposure, the sodium sulphate was replaced based on the change in the pH of the solution with a new batch.

3.3.4 Sodium Chloride

A condition representing an aggressive environment was set up using 5% sodium chloride (NaCl) solution added by weight of volume. The reagent-grade NaCl was mixed with distilled water to prepare the sodium chloride solution. The chemical properties of sodium chloride are presented in Table 3.2.

Table 3.2 Chemical analysis of sodium chloride

Compounds	Percentage composition (%)
Assay	99.8
Insoluble	0.02
Loss on drying	0.20
Iodide, I	0.012
Bromide, Br	0.10
Nitrate, NO_3	0.005
Sulphate, SO_4	0.005
Nitrogen compound, N	0.001
Phosphate, PO_4	0.001
Magnesium, Mg	0.005
Potassium, K	0.04
Calcium, Ca	0.01
Iron, Fe	0.0005
Arsenic, As	0.0001
Heavy metals, Pb	0.001

3.4 Compressive Strength Performance Under Tap Water

It is well known that the initial concrete strength could be attained from the first day up to 28 days; meanwhile, long-term strength is acquired after 28 days. In the present study, the compressive strength of SCC-BA specimens (100 × 100 × 100 mm) was tested at the age of 7, 28, 60, 90, and 180 days. The average of each three specimens was taken into account. The data of the compressive strength test on self-compacting concrete mixes are presented in Fig. 3.1.

Based on the results, concrete strength had increased with an increase in the ageing period. This could be due to the pozzolanic reaction that occurred in the concrete specimens. Similarly, the compressive strength increased with an increase in CBA content up to 10%. Beyond that, the compressive strength significantly decreased on further addition. For example, the compressive strength was found in the range of 58.56–68.25 MPa at the age of 28 days. In particular, the concrete mix containing 10% of CBA exhibited a higher 28-day strength compared to 30, 25, 20, 15, and 0% replacement of coal bottom ash concrete mix. At 15% replacement of CBA, the compressive strength was lower than the control strength. Thus, it could be inferred that the optimum replacement percentage of sand by CBA coal bottom ash was 10% in terms of the compressive strength of concrete exposed to tap water.

It was also observed that that the compressive strength of concrete incorporating CBA decreased significantly at an early age, while the compressive strength increased at a later age and exhibited higher strength than the reference specimens without bottom ash. This positive result probably can be attributed to the pore refinery effect by the pozzolanic reaction of CBA to enhance the formation of calcium aluminate

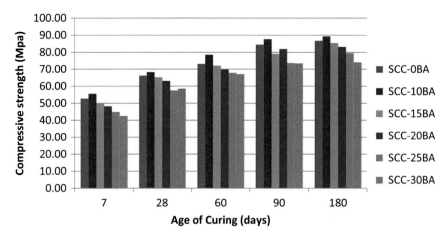

Fig. 3.1 Compressive strength of concrete cured in tap water

hydrate (CAH) and calcium silicate hydrate (CSH). It was found that the strength at 10% replacement was marginally higher compared to that of the control concrete. Therefore, it could be inferred that the optimum percentage of CBA replacement can densify the concrete pores and thus slightly increase the compressive strength.

3.5 Compressive Strength Performance Under Sodium Chloride Solution

Figure 3.2 presents the compressive strength of SSC incorporating CBA exposed to NaCl solution with wetting–drying cycles. It could be seen that the compressive strength of SSC specimens had increased up to 90 days. After that, it started to drop until the specimens finally deteriorated after 180 days. At the initial stage, the improvement of SCC strength resulted from the combined effect of the formation of ettringite needles, crystals of portlandite bonds, Friedel's salt, and the synergistic effect. Ettringite is a hydration product of cement predominantly formed as a result of C_3A reaction, which influences premature strength development.

On the other hand, the reduction in the strength of SCC was mainly due to two reasons. The reaction of cement hydrates with sodium chloride and severity level of the exposure system is regarded as the first reason that causes concrete deterioration. The second reason is related to the repetitive crystallization cycles of $NaCl \cdot H_2O$ by wetting–drying cycles, which would induce internal stresses in the pores and lead to cracks.

Sodium chloride or NaCl solution is a well-known aggressive chloride ion that reacts with calcium hydroxide in the binder paste to form gypsum ($CaSO_4 \cdot 2H_2O$). This reaction is related to an increase in the concrete volume. Apart from

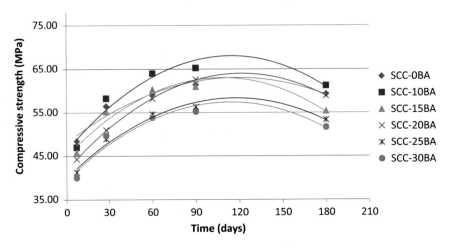

Fig. 3.2 Compressive strength of SSC exposed to NaCl solution

that, the destructive reaction between gypsum crystals and calcium aluminate in the binder paste has formed the less soluble reaction product, ettringite ($3CaO \cdot Al_2O_3 \cdot 3CaSO_4 \cdot 32H_2O$).

Overall, the strength of all mixes initially increases between 7 and 90 days; the increasing coal bottom ash content generally results in lower compressive strength, except for the 10% mixture. Incorporating a 10% replacement level of coal bottom ash increases the resistance of concrete against the chloride solution compared to the control. Figure 3.2 shows self-compacting concrete's performance with coal bottom ash as partial replacement to sand under chloride attack, which was monitored for up to 180 days. The mixtures of concrete cured in tap water did not show any significant difference in compressive strength from 7 to 180 days (Fig. 3.1), while specimens exposed to chloride solution with wetting–drying cycles showed lower compressive strength compared to the control specimens.

The compressive strength had increased for samples aged for 90 days and then slightly decreased for those aged 180 days. Since mixes involved the inclusion of CBA, it may be deduced that coal bottom ash affects the strength of development phenomenon that occurs at around 180 days period. Therefore, it can be concluded that the compressive strength of coal bottom ash self-compacting concrete exposed to chloride solution has increased compared to the control concrete. However, the extent of increase was less than that of the measured strength cured in tap water. For instance, for the 10% replacement specimens cured in water, the strengths recorded for 7, 28, 60, 90, and 180 days were 55.61, 68.24, 78.54, 87.73, and 88.01 MPa, respectively. However, when sodium chloride solution is exposed with wetting and drying cyclic, the reduction in strength at the same age was approximately 0.23–15.01%.

3.5.1 Impact of Chloride Attack on Compressive Strength Reduction

The reduction in compressive strength was considered and calculated to assess the influence of chloride solution on SSC-BA performance exposed to chloride attack. Three concrete specimens from chloride solution and tap water at each period were tested, and the average values were taken into account. The reduction in the compressive strength of specimens exposed to chloride solution with wetting and drying cycles is shown in Fig. 3.3. The reduction of concrete strength was observed to be dependent on the content of bottom ash. Increasing the percentage of sand replacement by CBA would reduce the compressive strength loss due to ettringite formation. The ettringite is dependent on the amount of $Ca(OH)_2$ and C_3A present in the concrete. As the pozzolanic reaction proceeds in binder paste, $Ca(OH)_2$ is consumed to react with silicon dioxide (SiO_2) and water.

In the same context, it was noticed that the SCC-30BA witnessed the highest reduction in compressive strength. In contrast, 10% replacement of CBA showed the lowest reduction in the compressive strength. In general, the strength reduction of SCC-30BA, SCC-25BA, SCC-20BA, SCC-15BA, SCC-10BA, and SCC-0BA concrete was about 2.36%, 2.05%, 2.04%, 1.83%, 1.13%, and 1.15%, respectively, at the age of 28 days. The compressive strength reduction of SCC-30BA, SCC-25BA, SCC-20BA, SCC-15BA, SCC-10BA, and SCC-0BA concrete was about 15.01%, 14.33%, 13.69%, 13.20%, 12.05%, and 12.92%, respectively, at the age of 180 days.

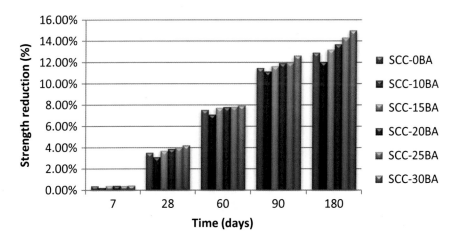

Fig. 3.3 Compressive strength reduction of SSC-BA exposed to chloride attack

3.5.2 Impact of Chloride Attack on Weight Loss

The weight loss of specimens was also considered to evaluate SCC-BA performance exposed to sodium chloride attack in wetting–drying. The weight loss of control specimen and coal bottom ash specimens of self-compacting concrete had primarily increased due to increased exposure time, as seen in Fig. 3.4. For example, it was found that the weight loss at the age of 28 days was 3.84%, 3.76%, 3.74%, 3.68%, 3.63%, and 3.66% for SCC-30BA, SCC-20BA, SCC-15BA, SCC-10BA, and SCC-0BA, respectively. At 180 days cycles, the weight loss of SCC-30BA, SCC-25BA, SCC-20BA, SCC-15BA, SCC-10BA, and SCC-0BA specimens were 15.20%, 14.47%, 14.55%, 14.06%, 13.16%, and 13.49%, respectively. This fact is attributed to the pores refinery effect by the expansive reaction products, thereby increasing the density of the hardened concrete mix. The increase in the pozzolanic reaction between coal bottom ash and $Ca(OH)_2$ will produce additional CSH. Consequently, this leads to an increase in the density and impermeability of the concrete.

3.6 Compressive Strength Performance Under Sodium Sulphate Solution

Figure 3.5 shows the graphical illustration of compressive strengths of SSC-BA exposed to 5% sodium sulphate solution with wetting–drying cycles at the age of 7, 28, 60, 90, and 180 days. It is interesting to note that the compressive strength of self-compacting concrete improved up to 90 days. After that, the compressive strength began to drop until the specimens finally deteriorated at the age of 180 days.

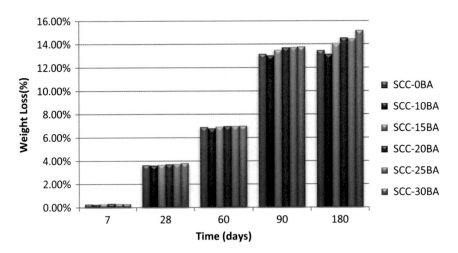

Fig. 3.4 Weight loss of SSC-BA exposed to chloride attack

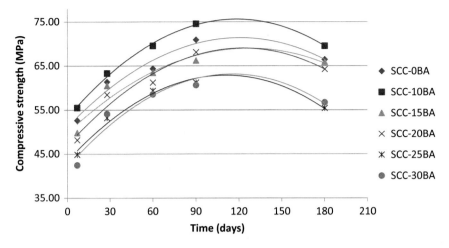

Fig. 3.5 Compressive strength of SCC-BA exposed to sulphate attack

The improvement of the compressive strength of SCC-BA exposed to seawater over time was attributed to the decrease in concrete pores. For example, prolonged hydration of unhydrated cement particles led to a gel product (calcium silicate hydrate) inside the concrete matrix. Similarly, in CBA, silica could react with calcium hydroxide to form an extra gel that can densify the concrete microstructure; however, the amount of silica is limited. Moreover, sulphate ions react with hydrated cement particles to form ettringite and gypsum, contributing to the increase in compressive strength. In other words, these chemical products lead to solid and denser structures as a result of precipitated products within the pores.

On the other hand, the decrease in the compressive strength observed in this laboratory experiment was predominantly due to the harshness of wetting–drying cycles. The sodium sulphate reacted with cement hydrates and monotonous sulphate crystallization by continual hydration process. The development of reactions has led to the formation of microcracks and resulted in a decrease in strength. It can be concluded that the strength of coal bottom ash self-compacting concrete subjected to 5% sodium sulphate with wetting–drying cycles were observed in three stages: (i) increased up to 28 days, (ii) linearly increased 28–90 days, and (iii) accelerating failure after 180 days.

In sodium sulphate exposure, the mixture of 25% replacement coal bottom ash to sand has produced low compressive strength compared to SCC-20BA, SCC-15BA, SCC-10BA, and SCC-0BA when subjected to sodium sulphate solution with wetting–drying cycles and the strength deteriorated after 90 days. The concrete specimen of SCC-30BA showed a slightly lower compressive strength compared to SCC-25BA for more prolonged exposure. In the same context, it was found that the SCC-10BA exposed to sulphate attack demonstrated a higher compressive strength compared to other mixture.

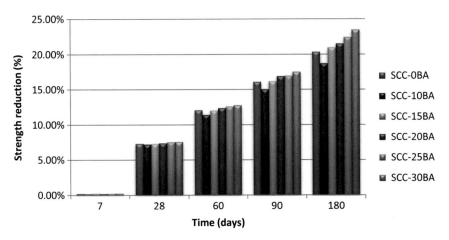

Fig. 3.6 Compressive strength reduction of self-compacting concrete incorporated with CBA and exposed to sulphate attack

In general, under wetting and drying cyclic tests, the compressive strength of SCC-BA samples exposed to sulphate solution experienced a declining trend compared to that of the SCC-BA samples exposed to tap water. For example, the extent of compressive strength reduction of SCC-30BA, SCC-25BA, SCC-20BA, SCC-15BA, SCC-10BA, and SCC-0BA was found to be 12.80%, 12.63%, 12.40%, 12.05%, 11.45%, and 12.10% after 28 days, respectively, while these values reached 23.51%, 22.46%, 21.55%, 21.00%, 18.75%, and 20.37% after 180 days, respectively.

It can be inferred that there is not much difference between the compressive strength of specimens exposed to the sulphate solution and the specimens subjected to sodium chloride solution. For example, under wetting–drying cycles, the compressive strength of SCC-10BA exposed to sulphate solution was found to be 63.31 MPa after 28 days (see Fig. 3.6). In comparison, the compressive strength of SCC-10BA exposed to sodium chloride solution was 58.25 MPa after 28 days. This indicated the differences in compressive strength of SCC-10BA exposed to sulphate and chloride solution, i.e. 7.99%.

3.6.1 Impact of Sulphate Attack on Compressive Strength Reduction

The influence of sulphate solution on the compressive strength of coal bottom ash self-compacting concrete was evaluated. The reduction in the strength of concrete specimens exposed to sodium sulphate solution with wetting–drying cycles is illustrated in Fig. 3.6. It was found that as time increased, reduction in the compressive strength also increased up to the age of 180 days. For example, the reduction in the compressive strength was found to be 7.58%, 7.54%, 7.38%, 7.31%, 7.22%, and

7.33% for SCC-30BA, SCC-25BA, SCC-20BA, SCC-15BA, SCC-10BA, and SCC-0BA after 28 days, respectively. Besides, the highest reduction in SCC-BA strength was recorded at the age of 180 days. This can be explained as $Ca(OH)_2$ is consumed during the reaction with silicon dioxide (SiO_2) accompanied by the formation of CSH and sulphate ions, resulting in the reduction of compressive strength while the pozzolanic reaction was progressing. Based on Fig. 3.6, the highest reduction in compressive strength was obtained for SCC-30BA, whereas SCC-10BA samples showed the lowest reduction in the compressive strength.

3.6.2 Impact of Sulphate Attack on Weight Loss

The weight loss of the SCC-BA specimens was also recorded and evaluated. In general, the weight loss of self-compacting concrete with coal bottom ash as fine aggregate replacement initially increased with the increase in the exposure time, as shown in Fig. 3.7. For instance, the weight loss due to sulphate attack at 28 days was 2.15%, 2.14%, 2.15%, 2.13%, 1.87%, and 2.01% for SCC-30BA, SCC-20BA, SCC-15BA, SCC-10BA, and SCC-0BA, respectively. However, at 180 days cycles, the weight loss of SCC-30BA, SCC-25BA, SCC-20BA, SCC-15BA, SCC-10BA, and SCC-0BA specimens were 14.83%, 14.43%, 14.13%, 13.57%, 12.41%, and 13.35%, respectively. The increasing loss of weight with exposure time may be due to the pores refinery effect that infiltrates the reaction products, thus increasing the density of the hardened concrete mix and decreasing the weight of specimens.

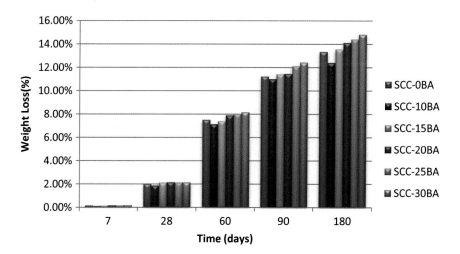

Fig. 3.7 Weight loss of self-compacting concrete containing CBA and exposed to sulphate attack

3.7   Rapid Chloride-Ion Permeability Test (RCPT)

Using ASTM C1202, the rapid chloride-ion permeability test (RCPT) was carried out to evaluate SCC-BA performance based on the charge passed. The total charge passed increased when the resistance of ingress chloride decreased. Figure 3.8 shows the performance of SCC-BA samples against chloride ion penetration with cyclic wetting–drying. With an increase in coal bottom ash content, the total charge passed of the concrete specimen also increased.

Generally, the value of the total charge passed of SCC-BA varied from each other based on the curing time and coal bottom ash content. The results agree with the findings from other researchers studied on the reduction in the total charge passed. Their study had involved the substitution of natural sand with additives. The high coal bottom ash contents will produce greater chloride permeability for a concrete age of 7 days for cyclic chloride exposure. It was also found that the chloride charge passed in all mixtures reduced with an increase in the concrete age. For instance, the total charge passed of SCC-10BA concrete at the age of 7, 28, 60, 90, and 180 days are 3510, 2510, 2111, 1820, and 1555 Coulombs, respectively.

The significant reduction in rapid chloride-ion permeability in self-compacting concrete may be due to the development of secondary calcium silicate hydrate due to the pozzolanic reaction, reducing the pore's size, making the hardened concrete denser, and therefore reducing the charge passed. The coal bottom ash replacement to the sand in concrete mixes can reduce the total charge passed even subjected to chloride solution with wetting–drying cycles. It can be concluded that the use of coal bottom ash meritoriously increased the resistance of hardened paste to chloride penetration at 10% replacement.

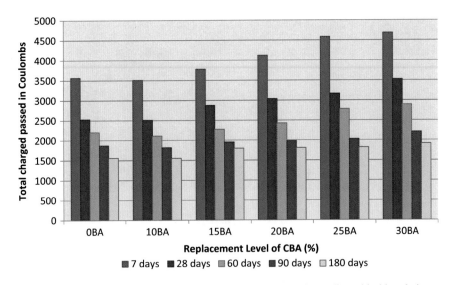

Fig. 3.8 Measurements of total charge passed of concrete exposed to sodium chloride solution

The ASTM C1202 standard established by the American Society for Testing and Materials recommends the approximate values to evaluate the RCPT results. The values of total charge passed were established at the end of 6 h of testing. The risk of penetrability of chloride ion is considered "high" if the total charge passed higher than 4000 Coulombs, whereas the values between 100 and 1000 Coulombs are considered a "very low" risk. Table 3.3 shows the chloride-ion penetrability based on the charge passed for self-compacting concrete specimens exposed to a chloride environment. It clearly indicates that with the increase in the replacement level, the total charge passed increases. Nevertheless, at 10% replacement coal bottom ash to sand, the RCPT results are slightly lower compared to control, whereas the 30% replacement specimen shows a higher total passed charge. This shows that the optimum replacement percentage of coal bottom ash is 10% when it is subjected to a chloride solution.

Meanwhile, for sulphate exposure, it was also found that the total charge passed decreased on increasing the age of curing (Fig. 3.9). For instance, at 28 days of exposure, the total charge passed of 10% replacement coal bottom ash to sand was 1893 Coulombs, whereas at 15%, 20%, 25%, and 30% replacement coal bottom, the corresponding total charge passed was 1941, 2002, 2317, and 2482 Coulombs, respectively. It has been found that 10% of replacement was recorded as the lowest charge passed, whereas 30% of replacement was the highest total charge passed from all observed percentages.

Coal bottom ash is a by-product of fine aggregates with low chloride permeability and high chloride binding capacity, although their finenesses and specific surfaces are not strong enough to bind more chloride ions. The lower permeability of concretes containing a 10% replacement ratio, compared to other concretes containing different replacement ratios, can be resulted from the pozzolanic reaction. The coal bottom ash reacted with free lime (CaO) during hydration and produced an additional substance with the silicate gels of cement. Therefore, it resulted in capillary pores of 10% with filled replacement coal bottom ash concrete. It was also found that the capillarity of the SCC-10BA specimen was lower when the replacement percentage increased due to the decrement of capillary pores. It is also suggested that the specimens containing higher replacement to sand ratios have higher capillarity due to porous structure composed by the usage of coal bottom ash in concrete.

In comparison, the total charge passed of specimens subjected to sulphate solution was the lowest compared to specimens subjected to the chloride solution. For instance, at 10% replacement coal bottom ash, the total charge passed of the specimen subjected to chloride solution at 180 days was about 1555 Coulombs compared to approximately 927 Coulombs observed in specimens subjected to sulphate solution, which indicates a corresponding reduction of 40.39%. From the data presented in Table 3.4, it can be observed that the total charge passed of concrete specimens varies from "high" to "very low" due to an increase in age ranging from 7 to 180 days.

Table 3.3 Evaluation of chloride ion penetrability of concrete exposed to sodium chloride solution

Mix	Total charge passed in coulombs and risk of penetration									
	7 days	Risk	28 days	Risk	60 days	Risk	90 days	Risk	180 days	Risk
SCC-0BA	3567	Moderate	2530	Moderate	2201	Moderate	1872	Low	1560	Low
SCC-10BA	3510	Moderate	2510	Moderate	2111	Moderate	1820	Low	1555	Low
SCC-15BA	3783	Moderate	2877	Moderate	2278	Moderate	1957	Low	1800	Low
SCC-20BA	4126	High	3039	Moderate	2431	Moderate	1983	Low	1811	Low
SCC-25BA	4589	High	3167	Moderate	2785	Moderate	2031	Moderate	1823	Low
SCC-30BA	4690	High	3525	Moderate	2894	Moderate	2210	Moderate	1920	Low

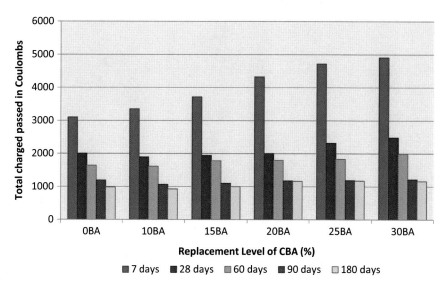

Fig. 3.9 Measurement of total charge passed of concrete subjected to sodium sulphate solution

3.8 Chloride Penetration by Rapid Migration Test

This section discusses the effect of coal bottom ash replacement to sand in self-compacting concrete subjected to chloride and sulphate solution. The characteristics of the resistance to chloride migration are also described. Chloride penetration by rapid migration test is measured by splitting the specimens after the test and spraying them with silver nitrate. The data on penetration rate for specimens exposed in chloride solution by wetting–drying cycles are illustrated in Fig. 3.10.

As shown in Fig. 3.10, it can be observed that the use of coal bottom ash as a partial replacement to sand has a significant influence on chloride resistance. The chloride penetration rate increased as the level of replacement of coal bottom ash increased. From the results, it can be observed that the chloride penetration rate of all mixtures reduced with the increase in the concrete age. Moreover, the mixture of SCC-10BA exhibited a lower chloride penetration rate compared to control for all the periods. For instance, at 28 days, the chloride penetration rate for SCC-0BA was 1.10 mm/(v.hr), while SCC-10BA, SCC-15BA, SCC-20BA, SCC-25Ba, and SCC-30BA achieved at rates of 0.90, 1.05, 1.35, 1.69, and 1.82 mm/(v.hr), respectively. The use of coal bottom ash as partial replacement of sand is limited to a replacement ratio of 10%, which is marked by decreasing the chloride penetration rate compared to the control mix. This is associated with the concrete capillarity of coal bottom ash concrete that caused lower penetration rates. Coal bottom ash in concrete produced a lower chloride penetrability rate and high chloride binding capacity. The coal bottom ash reacted with CaO during cement hydration and produced expansive products to the silicate gels of cement. As a result, the capillary pores of 10% replacement coal

Table 3.4 Evaluation of chloride ion penetrability of concrete exposed to sodium sulphate solution

Mix	Total charge passed in Coulombs											
	7 days	Risk	28 days	Risk	60 days	Risk	90 days	Risk	180 days	Risk
SCC-0BA	3102	Moderate	2001	Moderate	1638	Low	1193	Low	986	Very Low
SCC-10BA	3352	Moderate	1893	Low	1612	Low	1067	Low	927	Very Low
SCC-15BA	3718	Moderate	1941	Low	1783	Low	1101	Low	1002	Low
SCC-20BA	4325	High	2002	Moderate	1799	Low	1178	Low	1167	Low
SCC-25BA	4723	High	2317	Moderate	1836	Low	1192	Low	1179	Low
SCC-30BA	4911	High	2482	Moderate	1989	Low	1218	Low	1166	Low

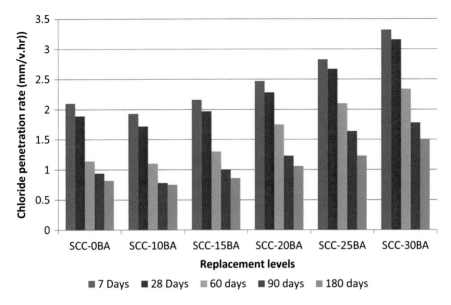

Fig. 3.10 Rate of the chloride penetration of concrete exposed to chloride solution

bottom ash concretes got filled. This simplified the reason for the chloride penetration rate to be increased with higher replacement coal bottom ash to sand in concrete. In addition, the chloride penetration rate at the age of 28 days decreased by 35% and then started to increase up to 180 days by 38% for the SCC-10BA mix.

The chloride penetration rates for specimens exposed to sulphate solution by wetting–drying cycles are shown in Fig. 3.11. The rapid migration test results on specimens subjected to sodium + sulphate solution through cyclic wetting–drying displayed that it was similar to those exposed to chloride solution. The use of coal bottom ash in concrete has a significant impact on chloride resistance.

The results showed that penetration rates increased with higher replacement of coal bottom ash. The 10% replacement of coal bottom ash mixture has exhibited the lowest chloride penetration rate compared to the SCC-0BA mixture. This may be attributed to the high chloride binding and concrete capillarity of coal bottom ash concrete that caused lower chloride penetration rates. During the hydration process, coal bottom ash has integrated with CaO compound and produced expansive products to the silicate gels of cement; thus, the concretes capillarity is packed. As a result, chloride penetration rate increased with higher replacement of coal bottom ash to sand in concrete. For instance, at an early age up to 7 days, the chloride penetration rates displayed the highest values for all mixtures. The chloride penetration rate for SCC-0BA was 1.52 mm/(v.hr), while SCC-10BA, SCC-15BA, SCC-20BA, SCC-25BA, and SCC-30BA achieved at rates of 1.39, 1.47, 1.73, 1.94, and 2.20 mm/(v.hr), respectively. The penetration rates considerably decreased on prolonging the exposure time to 180 days for all mixtures.

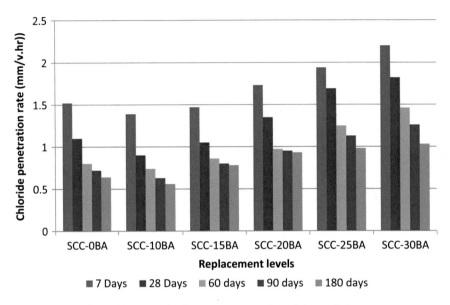

Fig. 3.11 Rate of chloride penetration in concrete exposed to sulphate solution

3.9 Carbonation Depth Test

Carbonation is a chemical reaction product, namely calcium carbonate ($CaCO_3$), occurring in the reaction between three main substances, which are water (H_2O), calcium phases (Ca), and carbon dioxide (CO_2). These three substances are attended to in different conditions; water is present in the concrete pores, calcium phases (mainly $Ca(OH)_2$ and CSH) are present in the concrete, and carbon dioxide is present in the surrounding air. In particular, carbon dioxide penetrates the concrete pores and dissolves to form carbonic acid (H_2CO_3) in the presence of water. Then, the carbonic acid reacts with the calcium ions available inside the concrete pores to produce the calcium carbonate. As soon as the $Ca(OH)_2$ is transformed and used up from the cement paste, the hydrated calcium silicate hydrate (CSH) releases calcium oxide (CaO) and will carbonate.

Figure 3.12 shows the evolution of carbonation depth of self-compacting concrete incorporating CBA when exposed to chloride solution. It could be seen that the carbonation depth decreased over time in all concrete mixes. This is attributed to the pore densification by the pozzolanic reaction of coal bottom ash at longer ages, particularly at 180 days of ageing. On the other hand, it can be observed that the carbonation resistance of 10% coal bottom ash replacement specimens was higher compared to the control specimens. Moreover, the carbonation resistance of all specimens decreased with the increase in coal bottom ash content. This can be attributed to the reduction in concentration elements of calcium hydroxide and calcium silica

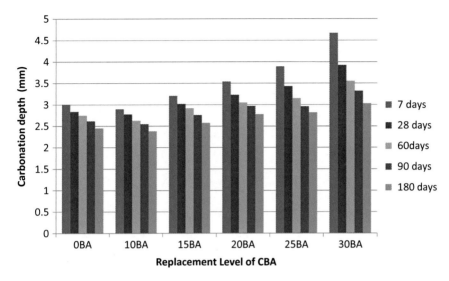

Fig. 3.12 Carbonation for SCC-BA exposed to chloride solution

hydroxide in concrete caused by substitution materials, leading to increased carbonation. Comparable results were reported by Turk (2012) on silica fume and fly ash in self-compacting concrete. For instance, at an early age up to 7 days, the carbonation depth displayed the highest values for all mixtures. The carbonation depth for SCC-0BA was 3.01 mm, while SCC-10BA, SCC-15BA, SCC-20BA, SCC-25BA, and SCC-30BA demonstrated rates of 2.90, 3.21, 3.54, 3.89, and 4.67 mm, respectively. The carbonation depth decreased significantly on increasing exposure time until 180 days for all mixtures.

The effect of the replacement of coal bottom ash in self-compacting concrete exposed to sulphate solution on carbonation depth is presented in Fig. 3.13. The carbonation test results on specimens subjected to sodium sulphate solution through cyclic wetting–drying displayed show similarity to those exposed to chloride solution. The carbonation depth of specimens decreased with time in all concrete mixes. This is also caused by the densification of concrete and pore size reduction when prolonging the concrete ages. In other words, the depth of carbonation decreases with increasing compressive strength for all types of concrete; however, this scenario is dependent on the type of cement and curing. Also, the carbonation depth of all specimens increased with an increase in the coal bottom ash replacement ratio. This can be described with an increase in the carbonation, which is caused by the reduction of calcium hydroxide $Ca(OH)_2$ and calcium silica hydroxide (CSH) compounds in concrete, wherein these compounds are produced by the reaction of hydrated cement paste in the presence of coal bottom ash. The pozzolanic reaction in coal bottom ash has consumed $Ca(OH)_2$, in which less amount of $Ca(OH)_2$ decreased the ettringite and gypsum formation. The $Ca(OH)_2$ is a corrosion-susceptible component in

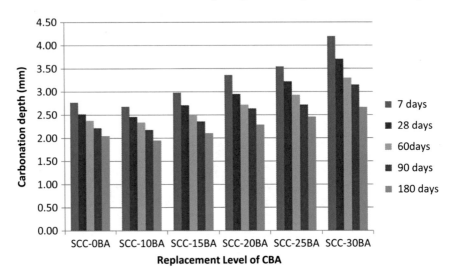

Fig. 3.13 Carbonation for SCC-BA exposed to sulphate solution

the concrete, wherein the resistance of concrete to sulphate attack improves with reduction in the $Ca(OH)_2$ amount.

It can be inferred that the carbonation resistance of 10% coal bottom ash replacement specimens was higher than that of control specimens. For instance, at 180 days, the carbonation depth displays the lowest values for all mixtures. The carbonation depth for SCC-0BA was 2.05 mm, while SCC-10BA, SCC-15BA, SCC-20BA, SCC-25BA, and SCC-30BA showed carbonation rates of 1.95, 2.11, 2.29, 2.46, and 2.67 mm, respectively.

3.10 X-Ray Diffraction (XRD)

The XRD analysis was conducted to identify the SCC-BA phase products exposed to chloride solution with wetting and drying cycles. Figure 3.14a shows the XDR of SCC-10BA exposed to sodium chloride solution up to 180 days. At $2\theta = 15°$ and 28°, the Friedel's salt peaks were detected. This result indicated that the chloride penetrated the concrete matrix and reacted with calcium aluminate hydrate to form Friedel's salt.

On the other hand, calcium silicate hydrate appears in diffractogram for all mixtures. The hydration of Portland cement is a chemical reaction to produce calcium silicate hydrate, which is responsible for the densification of concrete microstructure and thus increases the compressive strength of cement-based material. The quartz or silica oxide also appears in the diffractogram. About 57.43 and 62.57% of the total mass of coal bottom ash and fly ash were linked to the silica (SiO_2), respectively. The

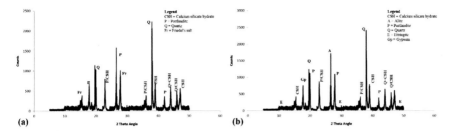

Fig. 3.14 XRD analysis for SCC-10% BA after 180 days of cyclic wetting–drying in **a** chloride solution, **b** sulphate solution

second constituent from both coal bottom ash and fly ash was found to be alumina (Al_2O_3).

Based on the analysis of XRD diffractograms, the calcium hydroxide product was also found in all mixtures. Also, the mixture without coal bottom ash replacement continuously produced portlandite throughout the hydration period.

In general, the presence of coal bottom ash in concrete could hinder the ingress of chloride ions into the concrete owing to the pozzolanic reactivity of CBA. The concentration of ions, such as H_3SiO_4, $Al(OH)_4$, and H_2SiO_4, also increased inside the concrete matrix due to the pozzolanic activity. The higher concentration of ions could disturb the chloride diffusion in concrete. Furthermore, coal bottom ash exhibited a higher content of alumina (Al_2O_3) compared to that of the fine natural aggregates. It was expected that the chloride penetrated in SCC-BA could be bound in the form of Friedel's salts.

On the other hand, sulphate solutions could deteriorate the cement-based materials through salt attack and crystallization. The sulphate attack may have occurred due to external factors such as penetration of sulphates, whereas the likely internal factors are the soluble sources being incorporated into concrete, such as gypsum in the aggregate. Besides, the negative effect of sulphate attack on the cement-based matrix was mainly linked to the formation of ettringite and gypsum. Specifically, sodium sulphate reacted with calcium hydroxide to produce gypsum. Gypsum, in turn, reacted with calcium aluminate hydrate to induce ettringite formation. The XRD analysis described the sulphate attack mechanism on self-compacting concrete and indicated that the compounds, such as portlandite, gypsum, and ettringite, were only crystalline phases present in the different depths of concrete in addition to quartz. For example, the diffractogram of self-compacting concrete containing 10% of coal bottom ash after 180 days of cyclic wetting–drying in 5% sodium sulphate solution is shown in Fig. 3.14b.

The peak corresponding to the ettringite formation was identified at 14°, 30°, and 47°. A high-intensity peak of gypsum was observed at 17°. On the other hand, the XRD patterns of specimens exposed to sodium sulphate solutions also showed a low-intensity thaumasite peak at 31°. The peaks of thaumasite and ettringite show high similarity due to the resemblance of their crystal structures. The peaks of portlandite, alite, quartz, and calcium silicate hydrate were detected at 21°, 26°, 37°, and 39°, respectively.

The XRD spectrums analysis revealed that the concentrations of ettringite and gypsum in SCC-BA were less than that of the control specimen. Moreover, the inclusion of coal bottom ash in SCC specimens might reduce the deterioration caused by sulphate attack, which could be attributed to the pozzolanic reaction in coal bottom ash. Calcium silicates hydrate (CSH) and portlandite formed from the reaction of free lime during cement hydration. It can be concluded that the XRD analysis indicates that the crystallization of sodium sulphate associated with the formation of ettringite has an extreme crystallization pressure compared to chloride solution.

3.11   Scanning Electron Microscopy (SEM)

Figure 3.15a, b shows the microstructure of self-compacting concrete containing 0% and 10% coal bottom exposed to the chloride environment obtained using SEM. The calcium silicate hydrate (C-S-H) gel was observed to be uniformly distributed inside

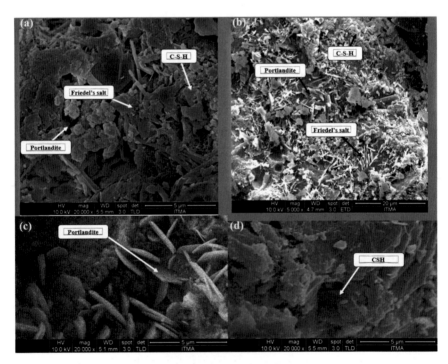

Fig. 3.15 Concrete microstructure **a** 0% CBA, **b** 10% CBA, **c** hexagonal-plate particle—portlandite crystal, **d** fine-layered particle—calcium silicate hydrate (CSH)

the concrete matrix. The other hydration products are Friedel's salt and other oxide minerals that naturally occur from calcium hydroxide, i.e. portlandite.

The morphology of portlandite crystal has a hexagonal-plate shape (Fig. 3.15c); meanwhile, the morphology of C-S-H shows a small encrusted particle (Fig. 3.15d) based on the SEM examination. Also, it could be seen that the intensity of the cement-based hydration production (C-S-H) of SCC-10BA was higher than the SCC-0BA mixture. In contrast, the amount of calcium hydroxide inside SCC-10BA was lesser than the SCC-0BA mix. This result was attributed to the pozzolanic activity inside the SCC-10BA mixture compared to that of the SCC-OBA mixture. Consequently, the hydration of cement could be improved in the presence of small particles of coal bottom ash that can decrease the volume of larger pores. As such, the concrete microstructure improved, which hindered the ingress of chloride ions and the carbonation process.

In the same context, the morphology of Friedel's salt found in self-compacting concrete is irregular shaped. Based on the micrographs, portlandite is predicted to decrease due to the pozzolanic reaction in the concrete mix. The mechanism can be explained when concrete components, including coal bottom ash, were mixed with water; the cement starts to hydrate and releases portlandite and calcium hydroxide crystal. Then calcium hydroxide reacted with coal bottom ash to produce calcium silicate hydrate, which will reduce the amount of calcium hydroxide in the solution, thus contributing to concrete strength development.

The microstructure of the samples that were exposed to sodium sulphate solution by wetting–drying at 180 days is presented in Fig. 3.16a, b. Based on the results, the main components of concrete microstructure included calcium silicate hydrate, calcium hydroxide (portlandite), calcium sulfoaluminate hydrate (ettringite and gypsum), sand, coarse aggregate, and ITZ between cement paste and aggregate.

The SEM analysis also conformed to the XRD results, wherein abundant formations of hydrated products were observed, such as ettringite and gypsum in ACC-BA exposed to sodium sulphate solution. The thaumasite crystal was also detected in micrographs (Fig. 3.16c). According to Nehdi and Bassuoni (2008), ettringite and thaumasite are usually grouped together in sulphate attack mechanisms. The ettringite appeared as needle-like or "club-shaped" crystals, whereas portlandite appeared as hexagonal platy crystals (Fig. 3.16d).

The morphology of gypsum was composed of "elongated rod-shaped" formations. The SEM results also revealed that incorporating coal bottom ash fine aggregate replacement decreased gypsum and ettringite content. This was due to the pozzolanic reactions responsible for reducing the amount of calcium hydroxide and reducing the further formation of ettringite and gypsum.

Overall, similar behaviour of deterioration was observed in both chloride and sulphate environments. However, the strength and hydration products produced in the sulphate environment are widely spread compared to the chloride environment, where there are large pores in the control specimen. The addition of coal bottom ash particles to fill these pores will in some way make the microstructure denser and stronger. The replacement of sand by CBA has significantly facilitated the resistance of concrete due to sulphate attack due to; mechanism of pozzolanic reactions; coal

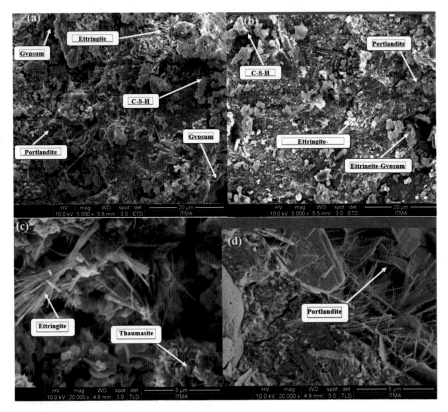

Fig. 3.16 Microstructure of concrete subjected to sodium sulphate solution **a** 0%, **b** 10% CBA, **c** ettringite and thaumasite, **d** hexagonal plate-like portlandite crystal

bottom ash in self-compacting concrete; and calcium hydroxide released during the cement hydration process.

3.12　Conclusion

Based on the obtained results, SCC-10BA samples exposed to both sulphate and chloride solution were found to be the optimum mixture in terms of compressive strength compared to that of the other mixtures. There was not much difference between the compressive strength of the SCC-10BA specimen exposed to the sulphate solution and the SCC-10BA specimens subjected to sodium chloride solution, in which the difference was found to be 7.99%.

In another positive scenario, the chloride penetration by rapid migration test, rapid chloride-ion permeability test, and carbonation test showed that the use of CBA had a positive impact on chloride attack resistance. In the chloride environment, the replacement of CBA in self-compacting concrete reduced chloride ion penetration by increasing the number of hydration products such as portlandite and Friedel's salt. Meanwhile, in the sulphate environment, the results showed that the replacement of sand by CBA significantly promoted concrete to resist a sulphate attack. This could be explained by the mechanism of pozzolanic reactions of coal bottom ash in self-compacting concrete and $Ca(OH)_2$ resulting from a cement hydration reaction, which reduces the amount of gypsum formed. Compared to the chloride and sulphate environment, cyclic exposure to chloride solution is slightly destructive to the sulphate environment. This has been confirmed by XRD and SEM analyses. Based on the inference made from both the characterization results, this study conclusively emphasized the coal bottom ash's role in tackling the durability aspect.

References

Jawahar, J. G., C. Sashidhar, I. R. Reddy, and J. A. Peter. 2012. A simple tool for self compacting concrete mix design. *International Journal of Advances in Engineering & Technology* 3: 550.

Turk, K. 2012. Viscosity and hardened properties of self-compacting mortars with binary and ternary cementitious blends of fly ash and silica fume. *Construction and Building Materials,* 37: 326–334.

Chapter 4
CBA Self-compacting Concrete Exposed to Seawater by Wetting and Drying Cycles

Abstract Concrete deterioration in a seawater environment has attracted considerable attention among concrete technologists. In the marine environment, sulphate and chloride ions are infiltrated into the concrete through seawater attacks. Therefore, self-compacting concrete with coal bottom ash was developed to resist the seawater attack and enhance the concrete's performance. The investigation of seawater attack on self-compacting concrete with CBA as a partial replacement of the fine aggregate is elucidated in this chapter.

4.1 Introduction

In the last few decades, studies on the concrete deterioration in the seawater environment have gained tremendous momentum among concrete technologists. In a marine environment, sulphate and chloride ions infiltrate into the concrete through seawater attack. Typically, sulphate ions will react with portlandite to form gypsum and will react with hydrated calcium aluminate to produce secondary ettringite. On the other hand, the chloride ions are the key parameters affecting the formation of Friedel's salt. Pozzolans such as fly ash, silica fume, rice husk ash, and slag are usually used in concrete production and have been widely incorporated in severe environmental conditions, such as marine environment and groundwater. However, investigations on the effect of supplementary cementitious materials such as coal bottom ash as a replacement to sand in seawater attacks with cyclic wetting–drying technique, especially in the local infrastructure in the context of Malaysia, are still limited. Furthering studies of seawater attacks on self-compacting concrete with coal bottom ash can provide extensive data for the design and maintenance of Malaysia's concrete infrastructures. In this chapter, the performance of self-compacting concrete incorporating coal bottom ash exposed to seawater with cyclic wetting–drying is explained. In the first section, a brief background on the mechanism of seawater attack is provided. Then, the proportions of self-compacting concrete incorporating CBA as well as curing conditions are also presented. Finally, the results of the mechanical and durability tests are also discussed in detail.

© The Author(s), under exclusive license to Springer Nature Singapore Pte Ltd. 2021
M. H. Bin Wan Ibrahim et al., *Properties of Self-Compacting Concrete with Coal Bottom Ash Under Aggressive Environments*, SpringerBriefs in Applied Sciences and Technology, https://doi.org/10.1007/978-981-16-2395-0_4

4.2   Mechanism of Seawater Attack on Concrete

The durability of the concrete structure exposed to seawater is a major concern, as seawater consists of two aggressive agents, namely sulphate and chloride ions. These ions are considered to be harmful substances for long-term durability. Besides, the tidal zone and the splash of repeated drying–wetting cycles also accelerate the deterioration of the concrete structure Chen et al. (2016).

It is interesting to note that the major problem in the marine environment is the sulphate attack. The concentration of sulphate in seawater is high and is usually present in the form of magnesium sulphate ($MgSO_4$) or sodium sulphate (Na_2SO_4). It, therefore, penetrates the concrete by capillary suction or diffusion. Magnesium sulphate has a negative impact on compressive strength, while expansive reactive products are formed by sodium sulphate. When both sulphate and chloride are present in seawater and penetrate the concrete, ettringite and Fridel's salt are produced, which results in concrete deterioration. The calcium silicate hydrate (C-S-H) phases and the calcium hydroxide are also affected by magnesium sulphate. Secondary brucite and gypsum result from the chemical reactions between sulphate, calcium hydroxide, and C-S-H. The weakness of the concrete structure and weight loss was due to the reaction of chloride to C-H of hydrated cement.

4.3   Self-compacting Concrete with CBA Mix Design

4.3.1   Seawater Sampling

The seawater sampling was conducted at the Malacca Strait in Pantai Punggur, Batu Pahat District, Johor Darul Takzim. The area was selected based on the availability of an open area and good wind circulation. The average monthly temperatures at this site range from 28 to 31 °C.

Based on the chemical analysis of seawater, the content of chloride and sulphate in seawater ranged from 16,000 to 19,000 ppm and from 2000 to 2500 ppm, respectively. Meanwhile, pH values of seawater were recorded at a 30 days interval. In the cyclic test, the specimens were subjected to an average of 15 h of wetting and 9 h of drying per day. The seawater in the laboratory was replaced every two weeks or as required based on the change in the pH.

4.3.2   Mix Concrete Proportion

A standard blended cement, 20 mm maximum coarse aggregate, and superplasticizer were used in all concrete mixtures to achieve the targeted strength of 40 MPa. The ratio of fine aggregate was 55% of the total aggregate weight. The mixture designs for self-compacting concrete were based on the method adopted by Jawahar et al. (2012). Three water–cement ratios were chosen to produce nomograph for self-compacting concrete with specific parameters that were tested earlier. For cyclic wetting and drying, a water–cement ratio of 0.40 was used to study the exposure of self-compacting concrete to an aggressive environment.

4.3.3 Samples Preparation and Curing Condition

To evaluate the performance of SCC-BA exposed to seawater, the proportions of the materials were first designed to obtain the target compressive strength of 40 MPa, as described in previous chapters. Then, the mixture was cast and cured in tap water for 28 days. After that, the SSC-BA specimens were subjected to seawater for different wetting–drying cycles.

For wetting cyclic conditions, the SCC-BA specimens were placed in a plastic container filled with 5% NaCl solution, 5% Na_2SO_4 solution, or seawater. The selection of the chemical concentration can cause severe damage to the tested mixture. It is considered to accelerate the results that resemble the field performance of concrete under harsh conditions within a reasonable testing period in the laboratory. All cyclic tests used identical 10 L plastic containers. For drying cycles, the SCC-BA specimens were kept in an enclosed chamber, and the temperature was in the range of (25–31 °C).

The duration of cyclic exposure was fixed at 15 h for the wetting cyclic, whereas the drying time was 9 h. This process was repeated until the exposure duration of both cyclic is completed. Five stages (7, 28, 60, 90, and 180 days) were taken into account to implement the compressive strength test of the SCC-BA samples.

4.4 Compressive Strength Performance Under Seawater

The results of the compressive strength of SCC-BA exposed to seawater are shown in Fig. 4.1. The general trend observed was practically comparable to sulphate and chloride solutions for up to 180 days of the exposure period. The strength seemed to increase up to 60 days. After 80 days, it started to decrease until the specimens had finally disintegrated. The involvement of two mechanisms could explain the increase in compressive strength. First, the chemical reaction between silica, which is present in the formation of coal bottom ash, and calcium hydroxide ($Ca(OH)_2$) occurred, resulting in producing extra gel (C-S-H). This positive result led to an increase in compressive strength due to the densification of concrete pores. Secondly, the compressive strength of SCC-BA exposed to seawater also improved due to the chemical reaction of the hydrate cement particles with chloride and sulphate ions formed by ettringite. The formation of ettringite also contributed to the increase in

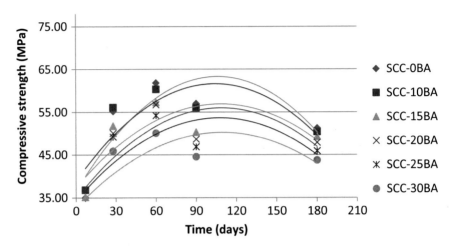

Fig. 4.1 Compressive strength of SCC-BA specimens exposed to sodium sulphate solution

the concrete strength. On the other hand, the decrease in compressive strength of SCC-BA can be attributed to two reasons. Firstly, disintegration was a result of the reaction between hydrated cement particles and magnesium sulphate. Second, the repeated wetting and drying cyclic promotes the continuation of the crystallization process ($MgSO_4.nH_2O$). As a result, cracks would be formed due to the stress generated inside the concrete pores. In general, the reaction of SCC-BA exposed to seawater is similar to that of chloride and sulphate solutions. The effect of seawater is attributed to the reaction of $Ca(OH)_2$ with $MgSO_4$ leading to the formation of $Mg(OH)_2$ and gypsum.

Based on Fig. 4.1, the SCC-10BA showed its benefit in resisting seawater attacks compared to other mixtures. This is because, for the ageing period of 28 days, the SCC-10BA exhibits a higher compressive strength compared to other concrete mixtures. However, the control mixture without the coal bottom ash has the highest compressive strength for a longer ageing period of 180 days, with a value of up to 51.23 MPa. While, the compressive strengths of SCC-30BA, SCC-25BA, SCC-20BA, SCC-15BA, and SCC-10BA were 43.72, 45.91, 47.89, 49.30, and 50.42 MPa, respectively. The compressive strengths of SCC-30BA, SCC-25BA, SCC-20BA, SCC-15BA, SCC-10BA, and SCC-0BA were 45.89, 49.20, 50.08, 51.79, 56.12, and 55.23 MPa at the age of 28 days, respectively.

Overall, the optimum replacement percentage of CBA in self-compacting concrete was found to be 10% compared to other mixture in terms of compressive strength even for different times of seawater exposure by the process of cyclic wetting–drying. The scenario of the incorporation of 10% of bottom ash in self-compacting concrete could be perceived as a sustainable and friendly strategy for the near future. Their potential benefits would dramatically increase compressive strength and reduce the cost of rehabilitation of affected concrete structures. They would also prolong the lifespan of concrete structures.

4.4.1 Impact of Seawater on Compressive Strength Reduction

The reductions in compressive strength of SCC-BA exposed to seawater environment were calculated and evaluated. From Fig. 4.2, it could be seen that with the increase in the ageing period, more reduction in compressive strength was observed. Moreover, on increasing the content of coal bottom ash in self-compacting concrete, the compressive strength decreased. The decrement of compressive strength of SCC-BA was started at the age of 28 days, attributed to the excessive amount of ettringite formation during that phase. For example, the reduction in compressive strength of SCC-30BA, SCC-25BA, SCC-20BA, SCC-15BA, SCC-10BA, and SCC-0BA were 2.36%, 2.05%, 2.04%, 1.83%, 1.13%, and 1.15%, at the age of 28 days, respectively. In addition, the reduction in compressive strength of SCC-30BA, SCC-20BA, SCC-15BA, SCC-10BA, and SCC-0BA were 39.28%, 38.52%, 36.28%, 34.64%, 31.54%, and 32.34%, respectively, after 180-day cycles. Based on the results, it was found that the exposure of repeated seawater wetting–drying has slightly affected the compressive strength of SCC-BA owing to the high amount of ettringite and gypsum as well as salts of crystallization. These products leach out and generate more pores inside the concrete matrix resulting in loss of strength.

Considering the above described scenario, it is inferred that as the amount of coal bottom ash replacement and time increases, the magnitude of compressive strength reduction also increases. The results are comparable with the findings of other researchers, wherein it was reported that the loss of strength was linked to the presence of an excessive amount of pozzolanic materials in hardened cement paste Zainal Abidin et al. (2015).

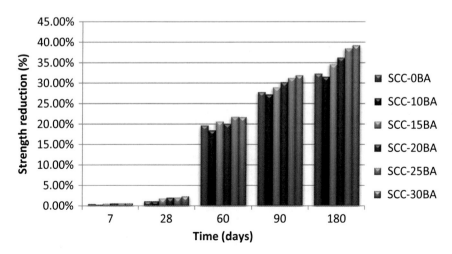

Fig. 4.2 Compressive strength reduction of SCC-BA due to seawater attack

4.4.2 *Impact of Seawater Attack on Weight Loss*

In the present section, the weight loss of SCC-BA is analysed and discussed. Figure 4.3 shows the weight losses of specimens exposed to seawater with wetting–drying cycles. After earlier age of exposure, the weight loss of both SCC-BA and control concrete initially increased. For example, after 28 days, the weight loss of control concrete specimens was 1.63%, followed by 1.72%, 1.70%, 1.68%, and 1.51% for SCC-30BA, SCC-20BA, SCC-15BA, and SCC-10BA, respectively. Moreover, the weight loss of SCC-30BA, SCC-25BA, SCC-20BA, SCC-15BA, SCC-10BA, and SCC-0BA samples were 4.92%, 4.58%, 4.69%, 4.28%, 4.03%, and 4.11%, respectively, at the age of 180 days.

4.5 Rapid Chloride-Ion Permeability Test (RCPT)

A comparison of the chloride-ion penetrability limits suggested by the ASTM C1202 standards for different concrete specimens is presented in Fig. 4.4. In general, the total charge passed of SCC-BA specimens subjected to seawater decreased when the exposure time increased. For instance, at 7 days of exposure, the total charge passed of SCC-10BA was 4001 Coulombs, while the total charge passed was recorded as 2854 Coulombs for 28 days of exposure. The decrease in total charge passed with age could be linked to the recurrent process of cement product hydration. The results also revealed that as the replacement level of fine aggregate by coal bottom ash increased, the total charge passed slightly increased. At 180 days, total charge passed value for SCC-30BA, SCC-25BA, SCC-20BA, SCC-15BA, SCC-10BA, and

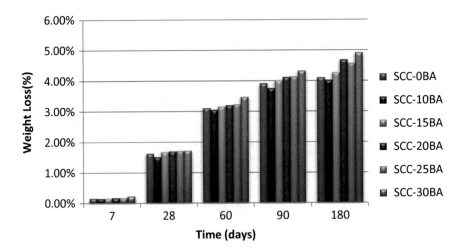

Fig. 4.3 Weight loss of SCC-BA due to seawater attack

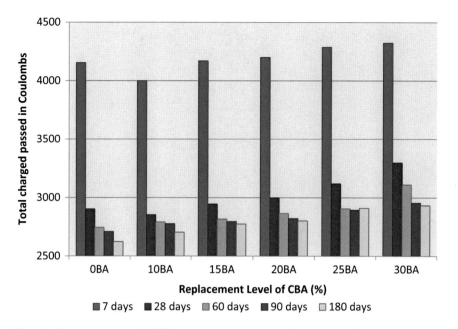

Fig. 4.4 Total charge passed SCC-BA subjected to seawater with wetting–drying cycles

SCC0-BA were 2934, 2911, 2802, 2774, 2704, and 2623 Coulombs, respectively. The capillarity mechanism of SCC-BA resulted in lower chloride-ion penetration. Coal bottom ash has a low degree of chloride permeability and high chloride binding capacity. However, the small particle sizes and specific surfaces of hardened cement are incapable of accumulating further chloride ions, resulting in low permeability.

The chloride-ion penetrability passed during exposure of the specimens to seawater with wetting–drying cycles is shown in Table 4.1. It can be inferred that the total charge passed of SCC-BA specimens varied from "high" to "very low" owing to age increment (from 7 to 180 days). Also, it can be noticed that the decrease in the chloride-ion permeability was related to the refinery process of concrete pore structure.

4.5.1 Relationship Between Compressive Strength and RCPT

The relationship between the rapid chloride-ion permeability test and compressive strength is plotted in Fig. 4.5. In this study, the specimens with a higher replacement ratio of coal bottom ash exhibited a higher value of total charge passed on being exposed to seawater by wetting–drying cycles. It is evident that the linear regression

Table 4.1 Chloride-ion penetrability passed of SCC-BA subjected to seawater with wetting–drying cycles

Mix	Total charge passed in coulombs											
	7 days	Risk	28 days	Risk	60 days	Risk	90 days	Risk	180 days	Risk
SCC-0BA	4156	High	2903	Moderate	2745	Moderate	2710	Moderate	2623	Low
SCC-10BA	4001	High	2854	Moderate	2791	Moderate	2777	Moderate	2704	Low
SCC-15BA	4172	High	2946	Moderate	2817	Moderate	2796	Moderate	2774	Low
SCC-20BA	4202	High	2999	Moderate	2865	Moderate	2821	Moderate	2802	Low
SCC-25BA	4289	High	3120	Moderate	2905	Moderate	2894	Moderate	2911	Low
SCC-30BA	4325	High	3298	Moderate	3111	Moderate	2956	Moderate	2934	Low

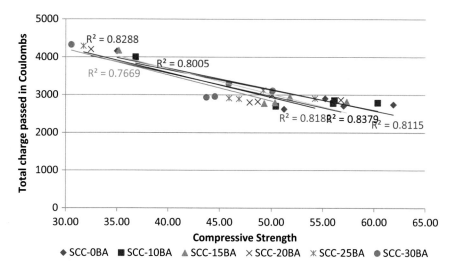

Fig. 4.5 Relationship between compressive strength and total charge passed of SCC-BA specimens subjected to seawater

of the different specimens is not significantly different. The inverse-type relationship shows the chloride-ion permeability decreases with an increase in compressive strength.

A strong correlation is observed between total charge passed and compressive strength. For instance, the total charge passed of the SCC-10BA mixture at 7 days is 4001 Coulombs (strength achieved is 36.81 MPa); while at 28 days, the total charge passed reduced to 2854 Coulombs with compressive strength of 56.12 MPa. Other mix types also demonstrated a similar trend with the replacement of CBA. The concrete mixture of SCC-15BA had the highest regression coefficient of 0.8379, while other mixtures showed a correlation value in the range of 0.76–0.84.

4.5.2 Relationship Between Time and Total Charge Passed

The relationship between total charge passed versus the time curve obtained through rapid chloride-ion permeability test is presented in Fig. 4.6. It was found that the correlation determination (R^2) was above 0.74, which indicates a good relationship between the total charge passed and the time. It is also evident that the chloride permeability of SCC-BA increased for a longer time, and a linear relationship was observed for each replacement mixture. This finding indicated that a specimen with a higher replacement ratio of CBA exhibited greater permeability of chloride ion when the total charge passed increased.

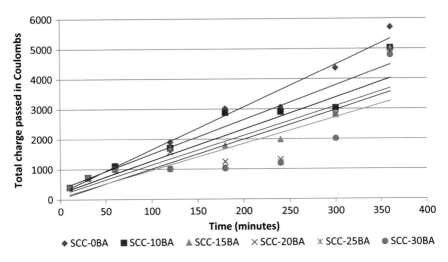

Fig. 4.6 Relationship between total charge passed and time (exposed to seawater)

The relationship between the current and total charge passed is shown in Fig. 4.7. The different percentages of coal bottom ash replacement mix show a linear relationship between the values with the coefficient above 0.78. It was observed that the increase in total charge passed indicates a higher current. Good linearity was also noticed between the total charge passed and the current with R^2 values exceeding 0.78. For SCC-10BA concrete, the total charge passed showed a coefficient of about 0.98 related to current.

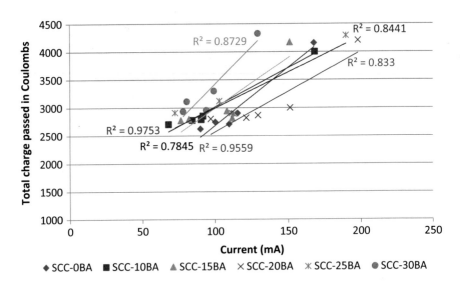

Fig. 4.7 Relationship between total charge passed and current (exposed to seawater)

4.6 Chloride Penetration by Rapid Migration Test

The chloride penetration test was performed to evaluate the performance of SCC-BA exposed to chloride attack according to ASTM C1202. After the period of exposure, the SCC-BA specimens were split and sprayed with silver nitrate. The result of the rapid migration test on SCC-BA specimens exposed to seawater has demonstrated that it was comparable to those open to chloride and sulphate solution. The chloride penetration rates for specimens exposed to seawater by repetitive wetting–drying are presented in Fig. 4.8. It can be observed that coal bottom ash has a major effect on the chloride-ion penetration in concrete.

At 180 days, maximum penetration rates were exhibited by all mixtures. Chloride penetration rates increased with an increase in CBA content in SCC. Coal bottom ash has a high chloride binding due to the porous texture and pozzolanic characteristics. Moreover, the lowest chloride penetration might occur due to the high chloride binding and capillarity of concrete. Hydration of cement paste controlled the creation of expansive products by reactions between coal bottom ash and silica material, thus producing silicate gels and making the concrete capillarity filled. Consequently, higher replacement levels incorporating coal bottom ash in concrete considerably increased the chloride-ion migration into concrete. As presented earlier, the chloride penetration rates showed the highest values at early ages for all mixtures. The SCC-30BA mixture achieved the highest chloride penetration rates compared to others. Meanwhile, SCC-10BA mixtures showed the lowest chloride penetration rates. The chloride penetration rate at 180 days for SCC-0BA was 0.82 mm/(v.hr), while SCC-30BA, SCC-25BA, SCC-20BA, SCC-15BA, and SCC-10BA achieved at rates of 1.5,

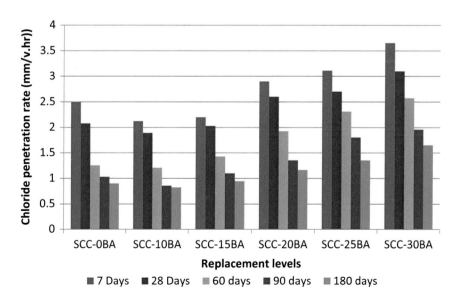

Fig. 4.8 Chloride penetration rate of SCC-BA exposed to seawater

1.23, 1.06, 0.86, and 0.75 mm/(v.hr), respectively. The good resistance of chloride-ion migration rates for specimens exposed to seawater occurred in a mixture of 10% coal bottom ash replacement.

4.6.1   Relationship Between Depth of Chloride Penetration and Total Charge Passed

The relationship between the depth of chloride penetration and total charge passed of self-compacting concrete exposed to seawater is presented in Fig. 4.9. As the chloride penetration depth increases, the total charge passed also increases based on the relationship plotted. For instance, the chloride penetration depth of the SCC-10BA mixture improved from 2.10 to 3.78 mm, which is attributed to the increase in the total charge passed from 2704 to 4156 Coulombs. Similarly, the chloride penetration depth increased from 2.71 to 4.82 mm due to the increment of the total charge passed from 2934 to 4325 Coulombs at the highest replacement ratio.

Furthermore, it was found that the specimens of SCC-10BA have the lowest total charge passed, characterized by a lower chloride penetration depth compared to the others. This significant finding was similar to the results of sulphate and chloride solution with cyclic wetting and drying. The trend can be attributed to the pore refinement by the pozzolanic reaction of coal CBA in concrete and alternate cycles of exposure. It also indicates that chloride-ion migration resistance decreased at the highest replacement ratio of coal bottom ash. When the concrete with high coal bottom ash has a high amount of calcium hydroxide, the chloride-ion migration will be disturbed. The concrete exposed to cyclic wetting–drying in the long run has

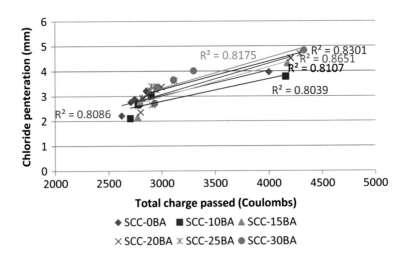

Fig. 4.9 Relationship between total charge passed and chloride penetration (subjected to seawater)

resulted in straightforward transportation of the chloride ions into the concrete. The relationship was described by linear regression with R^2 values above 0.80.

4.7 Carbonation Depth Test

Under wetting–drying cyclic, the performance of self-compacting concrete incorporating CBA exposed to seawater was evaluated using carbonation depth test, as shown in Fig. 4.10. According to the results, it was aligned to the concrete specimens subjected to sulphate and chloride solution. In particular, it was found that the carbonation depth decreased with an increase in time for all SCC-BA mixture. This positive fact has been attributed to the effect of pore refinery inside the concrete matrix.

It is also interesting to note that the increase in carbonation within the concrete matrix was associated with the decrease in the calcium hydroxide. In addition, the chemical reaction of the hydrated cement particles with the silica present in the coal bottom ash also led to the formation of other gel products, which, in turn, minimizes the porosity of the concrete. In other words, with an increase in the content of coal bottom ash in concrete, the extent of carbonation also increased. In the same context, the lower carbonation depth of the SCC-BA exposed to seawater was recorded when the bottom ash replacement percentage was 10%. It has been confirmed that the SSC-BA is more capable of resisting carbonation than any other mixture.

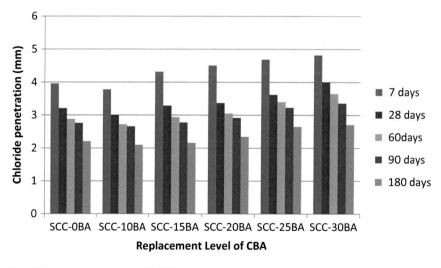

Fig. 4.10 Carbonation depth of SCC-BA exposed to seawater

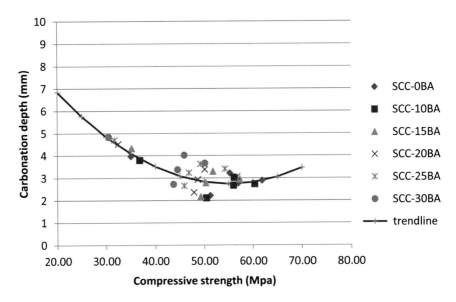

Fig. 4.11 Relationship between compressive strength and carbonation depth

4.7.1 Relationship Between Compressive Strength and Carbonation Depth

The relationship between the compressive strength of SCC-BA exposed to seawater and the carbonation test by the cyclic wetting–drying condition is shown in Fig. 4.11. The trends in SCC-BA exposure to seawater have shown similarity to other mixtures that have been subjected to chloride and sulphate solution. Based on the result, the highest carbonation depth measurement was associated with the lowest compressive strength. As such, it is inferred that the value of the carbonation depth is decreased by increasing the compressive strength. It is also important to note that the reduction in the compressive strength of SCC-BA exposed to seawater is attributed to the sulphuric acid attack. In particular, both sulphate and chloride ions present in seawater penetrate the concrete skin and form ettringite and Fridel's salt, which create weakness within the concrete matrix. The size of the SCC-BA samples would also be reduced due to the loss of cement paste, which would negatively affect the reduction of strength.

4.8 X-ray Diffraction

The X-ray diffraction (XRD) technique was utilized to identify the phases present in the hardened self-compacting concrete incorporating bottom ash subjected to seawater with wetting–drying cycles. The target powder cement paste of SCC-BA

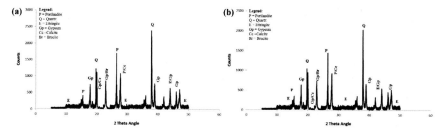

Fig. 4.12 Diffractogram of concrete after 180 days of cyclic wetting–drying in seawater **a** 0% CBA, **b** 10% CBA

specimens was tested at the age of 180 days. Based on Fig. 4.12a, the element of quartz and portlandite shows a relatively large peak intensity in control specimens exposed to seawater. The conversion of calcium hydroxide to gypsum and brucite was also evident from the diffraction patterns. The existence of calcite has revealed the carbonation effect on the concrete surface that reduces the coal bottom ash concrete (Fig. 4.12b). The ettringite phase was also identified for all specimens.

The main diffraction angle of ettringite was identified at 30° from the XRD spectrum. As it is well known, ettringite formation is often linked to the formation of gypsum, which would also have adverse and expansive effects. Specifically, ettringite is a chemical product obtained as a result of the reaction between calcium aluminate and gypsum. On the other hand, the intensity of brucite at 23° was notably observed in the concrete specimens. This is because the calcium hydroxide or portlandite could be reacted with magnesium sulphate to form gypsum and/or brucite. The diffractogram analyses also detected a small amount of gypsum and ettringite peaks in each specimen, which signifies that these compounds were the primary corrosion products. Nevertheless, the incorporation of CBA in self-compacting concrete as sand replacement material has minimized the concrete deterioration caused by seawater. It can be seen that the SCC-BA specimens exposed to seawater had lower intensity peaks of gypsum, brucite, and ettringite compared with control. Coal bottom ash had a pozzolanic characteristic, and it combines with cement to produce more silica dioxide than that of the control specimen. It is assumed that the materials incorporated produce CSH by consuming portlandite. In addition, the effect of CBA on portlandite composition also was significant. Consequently, the SCC-BA was protected against seawater attacks due to the decrease in the portlandite content in cement matrix.

4.9 Scanning Electron Microscopy

Figure 4.13a shows SEM analysis of SCC without coal bottom ash subjected to cyclic seawater. It was found that the self-compacting concrete contained irregular particles with micropores and cracking. These results provide insight into the need to use CBA to fill these pores and thus enhance both strength and durability. The

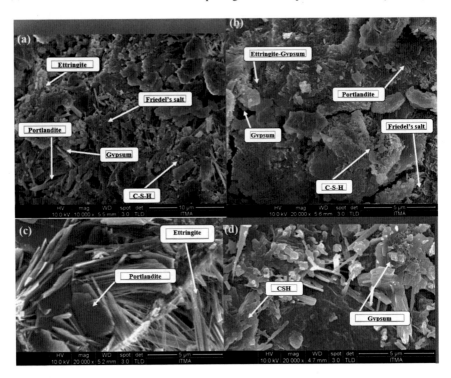

Fig. 4.13 Concrete microstructure subjected to seawater **a** 0% CBA, **b** 10% CBA, **c** ettringite and portlandite, **d** CSH and gypsum

micrographs of specimens subjected to seawater showed various microstructural formations, including Fridel's salt, ettringite, gypsum, and thaumasite.

As it is known, the seawater attack on the concrete specimen intensified owing to the chemical reaction of calcium hydroxide with magnesium sulphate, which later produced magnesium hydroxide and gypsum. Then, the calcium hydroxide reacted with the produced gypsum to form ettringite inside the cement paste. As illustrated in Fig. 4.13c, the needle-like crystals are ettringite, and portlandite is seen as hexagonal plates. Furthermore, C-S-H gel precipitated as a fine encrusted compound, and gypsum was formed in the form of an elongated rod-shaped formation (Fig. 4.13d).

Figure 4.13b shows the micrograph for SCC-10BA exposed to cyclic seawater. It was observed that the morphology of CSH crystal in the cement paste is controlled by the collective influence of chloride and sulphate ions contained in seawater. In this case, microcracks, as well as pores, were not observed. The C-S-H gel was uniformly distributed inside the SCC-BA matrix resulting in a highly dense structure and hindered the ingress of chloride ions into the concrete. This result was in line with the results obtained from the RCPT test in which SCC-BA showed its ability to resist harsh conditions such as chloride and sulphate attack.

4.10 Conclusion

The incorporation of CBA into SCC could be a sustainable strategy to improve the resistance to seawater attacks. The optimum replacement of fine aggregate by coal bottom ash was found to be 10%. Furthermore, it was observed that the increase in CBA replacement from 10 to 30% in SCC decreased the amount of gypsum and ettringite formation. Compared to chloride and sulphate exposure, the deterioration of SCC-BA specimens exposed to seawater was found to aggravate due to the presence of chloride and sulphate ions in seawater.

References

Chen, Y., J. Gao, I. Tang, and X. Li. 2016. Resistance of concrete against combined attack of chloride and sulfate under drying–wetting cycles. *Construction and Building Materials,* 106: 650–658.

Jawahar, J. G., C. Sashidhar, I. R. reddy, & J. A. Peter. 2012. A simple tool for self compacting concrete mix design. *International Journal of Advances in Engineering & Technology,* 3: 550.

Zainal Abidin, N. E., M. H. Wan Ibrahim, N. Jamaluddin, K. Kamaruddin, and A. F. Hamzah. 2015. The strength behavior of self-compacting concrete incorporating bottom ash as partial replacement to fine aggregate. *Applied Mechanics and Materials* trans tech publ: 916–922.

Printed in the United States
by Baker & Taylor Publisher Services